P9-AGQ-068

Darwin and the Nature of Species

SUNY series in Philosophy and Biology

David Edward Shaner, editor

Darwin and the Nature of Species

David N. Stamos

State University of New York Press

Published by
State University of New York Press, Albany
©2007 State University of New York

For information, address State University of New York Press,
194 Washington Avenue, Suite 305, Albany, NY 12210-2384

Production by Michael Haggett
Marketing by Susan M. Petrie

Library of Congress Cataloging-in-Publication Data

Stamos, David N., 1957-
    Darwin and the nature of species / David N. Stamos.
      p. cm. −
(SUNY series in philosophy and biology)
    Includes bibliographical references (p.   ) and index.
    ISBN-13: 978-0-7914-6937-8 (hardcover : alk. paper)
    ISBN-10: 0-7914-6937-9 (hardcover : alk. paper
    ISBN-13: 978-0-7914-6938-5 (pbk. : alk paper)
    ISBN-10: 0-7914-6938-7 (pbk. : alk. paper)
  1. Species–Philosophy.
  2. Darwin, Charles, 1809-1882. I. Title. II. Series.

QH83.S748   2007
578'.012–dc22

                                                              2005036225

10 9 8 7 6 5 4 3 2 1

*In memory of my mentor and friend the late Robert H. Haynes,
who enjoyed to the last what he called "the opiate of Darwinism."*

Contents

Preface

Looking back, I think it was more difficult to
see what the problems were than to solve them.
———Charles Darwin (letter to Charles Lyell, September 30, 1859)

The year 1859 marks the beginning of an enormous earthquake, an earth-quake that shook the world and continues to shake it to this very day. The earthquake and the consequent tremors were not caused by the gradual shift and strain of conflicting ideas, but by a sudden impact, the publica-tion of *On the Origin of Species* by Charles Darwin. It started a revolution in thinking, an enormous paradigm shift, the implications of which are still being worked out. Interestingly, at the very core of that revolution is the concept of species. It is important, then, to know exactly what Darwin did with that concept. The problem, however, is that for a variety of reasons scholars (biologists, philosophers of biology, and professional historians of biology) have provided interpretations that just don't fit the facts. A large part of the reason, as we shall see, was caused by Darwin himself. At any rate, the problem of Darwin on the nature of species, what was the prevail-ing view and how he tried to change that view, has yet to be adequately understood and appreciated. The time is definitely overdue for a detailed historical reconstruction. This becomes even more important because the concept of species in biology, from the time of Darwin right up to today, is still far from settled.

The purpose of this book is basically fourfold: First and foremost, to provide a full and detailed reconstruction of Darwin's species concept fo-cusing mainly on his mature evolutionary period, to get it right inasmuch as that is possible. In fact the present work breaks entirely new ground and constitutes a major reinterpretation of Darwin on the nature of species, in stark contrast to the literature on this topic, which stretches back over 140 years. Second, to apply Darwin's insights on the ontology of species to the modern species problem. Third, to take my reconstruction work on Darwin and apply it as a case study to a core issue in philosophy of science, namely, the problem of concept change in scientific revolutions. Fourth and finally, to use Darwin's species concept as an indictment against a now dominant trend in professional history of science.

I shall expand on these purposes later in this Preface, but first I want to deal with some preliminary matters, beginning with the identity of the specific audiences for which this book was written. Obviously it should be of great interest to Darwin scholars. They alone will be able to fully appreciate and enjoy the detailed historical work (even though it is *mainly* internalist) and the new direction that it takes. In fact anyone who is interested in things Darwinian should find this book worth their while. Historians of science, of course, should be especially interested not only for the work on Darwin but also for my application of it to what I call in the final chapter "the new historiography." The second major audience is biologists and philosophers who are interested in the modern species problem. In fact this book serves actually as a *prequel* to my previous monograph, aptly titled *The Species Problem* (2003), which focuses mainly on the modern species problem (the problem of determining whether species are real, and if real the nature of their reality; hence the problem of defining the species category). Darwin has much to say that is both interesting and important on this matter, although it has been almost entirely lost on subsequent scholars. Philosophers of science should also be interested, for the reconstruction work in the present book proves to be an enlightening case study for the topic of concept change in scientific revolutions, so much so that it presents a serious challenge to what many consider the received view.

The problem begins primarily with Darwin's most famous book, the full title of which is *On the Origin of Species by Means of Natural Selection, Or the Preservation of Favoured Races in the Struggle for Life* (Darwin 1859). In spite of the realist tone of its main title, Darwin repeatedly defines species, both individually and as a category, nominalistically, as arbitrary groupings and therefore as extramentally unreal. This is possibly the greatest enigma in the history of biology, even of science. Could it be that one of the greatest scientific minds of all time, and the main force behind what is arguably the most important scientific revolution of all time, was simply muddled on so basic an issue? The sheer irony is that for over a hundred years virtually everyone took him at his word, as believing that species are not real. Then in 1969 a major breakthrough was made by the biologist Michael Ghiselin, in his book *The Triumph of the Darwinian Method*

(1969). Ghiselin argued that species taxa for Darwin are real, such as *Canis lupus* and *Homo sapiens*, but not the species category, the class of species taxa and the object of a species definition. Sixteen years later John Beatty (1985) added to Ghiselin's thesis a strategy theory to explain why Darwin would define species nominalistically and yet hold that species taxa are real. For Beatty, Darwin simply followed the species designations of his fellow naturalists, but denied that the species category could be defined, simply to better communicate his evolutionary views, given that his audience had a theory-laden definition of "species."

Beatty's theory has enjoyed the status of being the received view ever since. The present book, on the other hand, is the first major-length study of Darwin on the nature of species, and one of its themes is that the received view should be received no more. Darwin did not simply follow the species designations of his fellow naturalists. Moreover the places where he declined to do so provide major evidence (in addition to other evidence) for reconstructing his implicit species concept.

Granted, there have been a few who have attributed a species concept to Darwin (e.g., a morphological species concept, or one involving sterility to some partial degree), implying a poverty of thought on Darwin's part, but in each case they succeeded only in revealing the poverty of their own research. The time is long overdue for a thorough and detailed analysis of Darwin's writings to bring out not only his actual species concept but also the richness and fruitfulness of that concept.

Although I make no claim to providing the last word on the subject, I do claim that this book marks a substantial advance. And it was not produced lightly. Rather it is the culmination of a period of research spanning roughly twelve years, parts of which have been presented in a number of publications (Stamos 1996, 1998, 1999, 2000, 2002, 2003, 2005), but most of which is new, the remainder being either completely rethought or refined.

When doing research on Darwin, I look at his writings much as a paleontologist looks at strata. Stephen Jay Gould (1989) put the case of the paleontologist best: "We search for repeated pattern, shown by evidence so abundant and so diverse that no other coordinating interpretation could stand, even though any item, taken separately, would not provide conclusive proof" (282). To perceive patterns that everyone else has missed, or to provide new and revolutionary interpretations for already perceived patterns, is the glory of the paleontologist (aside, of course, from discovering new bones). Although himself not a paleontologist, one has to think of the discovery by the physicist and Nobel laureate Luis Alvarez and his team and their explanatory theory, namely, the discovery of high levels of iridium at the K/T boundary and their theory of extraterrestrial impact to explain both it and the K/T extinction. The discovery of shocked quartz a few years later provided further considerable evidence in support. Interestingly, all of this had been missed by professional paleontologists, but it has now become the dominant theory in explanation of the mass extinction that leveled the dinosaurs and many other species 65 million years ago.

The same can happen in professional history. In the case of Darwin, the strata is the enormous amount of writings he left behind: his published books and articles, his manuscripts and notebooks, his correspondence and marginalia. Here there are still new patterns to be discovered and room for better theories to explain already discovered patterns. And one need not be a professional historian to do this. In fact, expertise in a different discipline, as with the Alvarez example, might be just what is needed to see what everyone else has missed and to thereby effect a paradigm shift. I make my case in the following chapters.

I shall also, as I've stated above, apply Darwin's insights to the modern species problem. But just what is that problem? Quite simply, it is the problem of determining the ontological status of species taxa, whether they are arbitrary mental constructs or real entities existing outside of the mind, whether they are something we make or something we discover; and if the latter is the case, it is the further problem of determining their precise nature and of formulating it in a definition.

The modern species problem has both purely theoretical and eminently practical dimensions. Beginning with the former, the species problem can be seen as the central problem of the Modern Synthesis. Begun in the 1920s with the marriage of Darwinian natural selection to Mendelian genetics, the union of the various subdisciplines in biology, ostensibly completed in the 1950s, has so far been without a unified species concept. Within each subdiscipline there are various contenders, and between them there has yet to be a clear winner. In fact, in the past three decades species concepts have proliferated in a Darwinian bush pattern, as evidenced by the recent anthologies devoted to the topics of species concepts and mechanisms of speciation (Otte and Endler 1989; Ereshefsky 1992a; Claridge *et al.* 1997; Howard and Berlocher 1998; Wilson 1999; Wheeler and Meier 2000).

The reason for this proliferation is not only theoretical. In addition to being the basal units of taxonomy and (as most think) the main units of evolution, species are also the main units of biodiversity. Unfortunately our world is in the midst of a major crisis in biodiversity, mass extinction #6 (mass extinction #5 being the one that occurred roughly 65 million years ago). The main cause of the current mass extinction, of course, is not extraterrestrial, but rather the rapid overpopulation of *Homo sapiens* and the corresponding destruction of the environment. According to the best estimate of Edward Wilson (1992, 278–280), we are losing 50–100 species per day and at the present rate shall reach roughly a 50% loss in biodiversity by the year 2050.

In response to this crisis, there has risen in recent decades a noticeable conservation movement, involving many different countries and levels of society, from grassroots to the United Nations. The main problem with laws and treaties is that (aside from the need for much greater funds) they need a unified species concept if they are to be uniformly applied. The situation is the same in other areas of law. Without an agreeable definition of pornography, for example, pornography laws cannot help but be vague or ambiguous and will accordingly suffer in their application.

The official species concept of the U.S. Endangered Species Act of 1973 explicitly employs in its definition of "species" the biological species concept, which is based on reproductive isolation and which was named by its most vociferous advocate, Ernst Mayr. Unfortunately this Act was made at a time near the end of the hegemony (at least in zoology) of the biological species concept. The current situation in biology is clearly that of pluralism, in that there are many species concepts actually in use in biology, and many more vying for contention. Some biologists, as we shall see in the next chapter, are species nominalists. Seemingly more are pluralists outright, believing that modern biology positively needs a variety of different species concepts to suit the needs of different biologists. Many have despaired of the species problem altogether and along with Robert O'Hara (1993) think that, like a dissolved marriage, we should try to "get over" (232) the species problem and simply get on with doing biology. Unfortunately this will not make the species problem go away.

Part of the problem is that different species concepts divide up the biological world in different ways. For example, Joel Cracraft (1997, 331) estimates that his phylogenetic species concept roughly doubles the 9,000 or so species of birds currently recognized by the biological species concept. More specifically a similar problem surrounds the case of the red wolf (Wayne and Gittleman 1995), the flagship of the U.S. conservation program. Millions of dollars were spent by the government, capturing, breeding, and reintroducing this species into the wild. Although an ambiguous species from the viewpoint of the biological species concept, it is a good species from the viewpoint of the morphological species concept. Recent DNA studies, however, have confirmed that the red wolf is a hybrid of the gray wolf and the coyote, and thus not a good species from the viewpoint of either the biological or phylogenetic species concepts. And yet from the viewpoint of an ecological species concept the government's efforts have been well spent.

Because of the biological species concept's lack of finer discriminations and other problems (including hybridization and its inapplicability to asexual forms), more and more biologists have been arguing that biology needs a better species concept, one that is universal in scope and that fits the needs of conservation biology. Indeed many still hold out hope for a universal species concept, one that will complete the Modern Synthesis and satisfy the various needs of biology, including conservation.

Now what has all of this to do with Darwin? I am certainly not under the illusion of thinking that whatever insights can be gleaned from Darwin's writings will be sufficient to solve the modern species problem. But I do think that his insights are sufficient to put us on the right track (and we are not on the right track!). As James Mallet (1995) put it,

> by 1859 he was an experienced systematist, having just finished his barnacle monograph, and had accumulated an encyclopaedic knowledge about species, both from his own travels and researches, and through prodigious correspondence with other zoologists and botanists. His private income left him free of bureaucracy and teaching; he had the time, the facts at his disposal, and the intellect to solve the problem of the nature of species. It is at least worthwhile reexamining Darwin's arguments. [294]

Mallet characterizes Darwin's species concept as "materialistic, morphological" (294). We shall see in later chapters that this is not at all close, and indeed that thus far no one else has come close either. But Mallet's point about Darwin's unique position and superior competence remains. In fact what we shall see in chapter after chapter is that Darwin had a wealth of insights highly relevant to the modern debate on species, insights that for the most part have gone unrecognized by virtually everyone who has written on the topic.

But surely, one might reply, even if we grant Darwin's unique position and superior competence, the situation was far different in Darwin's day compared to today. To a large extent, of course, this is true. Although there were many different species concepts bandied about in Darwin's day, the species problem was quite different from that of today. For a start, the main species concepts back then were at bottom creationist species concepts and the main issue was whether species are fixed or evolve. As Darwin put it in a letter to Leonard Jenyns (October 12, 1844) with regard to "the question of what are species," "The general conclusion at which I have slowly been driven from a directly opposite conviction is that species are mutable & that allied species are co-descendants of common stocks" (Burkhardt and Smith 1987, 67).

Today, of course, evolution is taken for granted, as a fact, so much so that "Nothing in biology makes sense except in the light of evolution" (Dobzhansky 1973). Accordingly the species problem has taken on quite a different meaning since Darwin's day. Although there are many issues that define the modern species problem, there are six in particular that I shall focus on in this book: (i) whether species are extramentally real or unreal (nominalism), (ii) whether species are abstract classes or concrete individuals, (iii) whether species are primarily horizontal or vertical entities, (iv) whether species can have multiple origins (polyphyletic) or must have single origins (monophyletic), (v) whether species are primarily process or pattern entities, and (vi) whether species must be consistent with history reconstruction. According to the botanist Melissa Luckow (1995), "the species problem will be solved by the continued collection and analysis of data, the clarification of issues and terms, and the application of new ideas" (600). Darwin, I shall argue, has something vitally interesting and important to say on each of the six issues just outlined, so much so that although his insights do not provide a final solution to the modern species problem they certainly help show us the way.

In chapter 1 I do not begin with reconstructing Darwin's species concept but instead spend most of the chapter providing a short history of nominalist interpretation of Darwin on species. Part of the problem, as we shall see right from the start, was created by Darwin himself. We shall also see, however, that Darwin, throughout the entirety of his career as an evolutionist, did indeed think that species are real.

In chapter 2 the reconstruction begins. What we shall see is that Darwin did have a distinction between species as a taxon and species as a category. Moreover, he provided a number of laws of nature for the species category, not for any particular species taxon but for the class of species taxa as a whole. Given that a number of thinkers on the modern species problem argue that the species category is objectively real because there are laws of nature that apply to it, I similarly argue that the species category (not just species taxa) was likewise objectively real for Darwin.

In chapter 3 I argue that the evidence is overwhelming that Darwin, to use a modern distinction, conceived of species as primarily (though not exclusively) horizontal entities in the Tree of Life. This is the first major step in understanding Darwin's view on the nature of species taxa. We shall also see that Darwin early in his career as an evolutionist toyed with but then rejected the idea that species are or are like individual organisms (which are temporally vertical entities), and that he did this in favor of the many analogies between species and languages, the latter of which for Darwin, as today, were thought of mainly as horizontal entities.

In chapter 4 I focus on Darwin's emphasis on common descent for natural classification. I show that in spite of his emphasis, Darwin did not subscribe to a concept of monophyly for species taxa (as is common today). Darwin's comments on extinction are relevant here, and my conclusion that Darwin did not insist on monophyly for species proves consistent not only with chapter 3 on the temporal dimension of species but also with what we shall see in chapter 5.

Chapter 5 presents the key to understanding Darwin's species concept, the key that everyone else has missed. What we shall find is that the key was not morphological discreteness, or characters constant and distinct, but adaptations. We shall see this in example after example in Darwin's writings, and it is the sole reason for why he went against his fellow naturalists (when he did) in their species designations. What we shall also see is that for Darwin adaptations were the key for distinguishing species not only because adaptations were the most amazing features of species, but because they were produced by a natural law, namely, what Darwin called "natural selection." Moreover, it will be shown that by bringing species under natural law, and also by using natural law to distinguish species, Darwin in one stroke was attempting to bring both species classification into the realm of scientific classification (which at that time put a high premium on natural law) and biology into the unity of science.

In chapter 6 I examine what was not part of the nature of species on Darwin's view, namely, reproductive isolation between species, fertility within a species, and the occupation of an ecological niche. Of particular interest are the reasons Darwin gave for rejecting these criteria, the first two of which enjoyed common currency in his own day and all three of which enjoy widespread currency today. Darwin's rejection of these criteria will also be seen to fit exactly with the criteria shown in previous chapters that he did accept. In short, Darwin clearly thought of species as pattern entities, not as process entities, and accordingly it is at the end of this chapter that a formal definition of Darwin's implicit species concept is given.

In chapter 7 I turn to a related issue, namely, Darwin's concept of variety. If Darwin thought that species are real and varieties are incipient species, then he must also have thought that varieties are in some sense real as well. Darwin's concept of variety has been even less explored than his concept of species, and this chapter attempts to make up for that glaring omission. For reasons given in the chapter, the various concepts of variety of Darwin's fellow naturalists are also explored. This is an area of research that has received pathetically little attention in the literature, and this chapter attempts to make up for it.

In chapter 8 I develop my own theory for why Darwin in the *Origin* and elsewhere explicitly denied that species (both category and taxa) are real and yet gave numerous indications elsewhere that he thought they are real. The theories of Ghiselin (1969), Mayr (1982), Beatty (1985), Hodge (1987), McOuat (1996, 2001), and my former self (Stamos 1996) are examined in detail and rejected, before presenting in detail what I believe to be the true pattern of Darwin's *modus operandi*.

In chapter 9 I broaden my focus and examine one of the basic issues in history and philosophy of science, namely, the problem of correctly modeling the nature of concept change in scientific revolutions. Beatty's (1985) interpretation of Darwin on species, which quickly became the received view, was, we shall see, influenced by a preconceived model of such change. Since his interpretation of Darwin on species does not fit the evidence, his failure raises anew the problem of a historically correct model. John Dewey's (1910) famous *get over it* model, we shall see, fails to refer, as well as the highly influential *incommensurability* thesis of Thomas Kuhn (1970, 1977), while Philip Kitcher's (1978, 1993) model of *reference potential* receives surprising corroboration (for both "species" and "variety"), even though he did not recognize it because (like so many others) he followed Beatty's (1985) strategy theory.

Chapter 10 shifts focus to an even more basic issue in history of science, namely, the issue of historiography. The current trend in professional history of science is externalist (the sociology of ideas), embedding scientists like Darwin in their time and place and keeping them there, in opposition to (even so much as trying to replace) the previous trend which was internalist (the history of ideas). The book presented here, of course, is not meant to be a competitor to biographies of Darwin, such as Desmond and Moore's (1992) controversial though highly influential biography. Nevertheless it is definitely meant to be a counterbalance to the modern trend in historiography, typified by Desmond and Moore's book. Indeed this final chapter, which builds on the chapters that precede it, serves as an indictment against that trend. To accomplish that task, I keep my eye on Darwin on the nature of species and respond not only to Desmond and Moore but to a number of other professional historians of biology who either try to denigrate the Darwinian revolution or Darwin's role in that revolution.

Although science is a human activity and all scientists work in a social context, when viewed from a bird's-eye view it should be clear that science does, to varying degrees, transcend its social milieu. Science is not merely a social construction. There are genuine revolutions in science and genuine progress in science, in an objective, epistemic sense, unlike, for example, the history of music. Moreover there are individual scientists who transcend not only their social milieu but also the science of their time. There is no finer example of this than Darwin. This is the grand theme of Michael Ghiselin's *The Triumph of the Darwinian Method* (1969), which I highly recommend. Although dated from the viewpoint of modern historiography, his book is still essential reading and is, on my view, basically correct. The present book extends Ghiselin's thesis in the one area in which it is truly outdated, namely, the issue of Darwin on the nature of species. Although Ghiselin did not get it right, he did nevertheless take great strides in the right direction. My ultimate thesis is that with his species concept Darwin belongs in the present time much more than his own, so much so that he still has plenty to teach us.

Acknowledgments

Special thanks to Editor in Chief Jane Bunker, Series Editor David Shaner, the two anonymous readers for SUNY, Michael Ghiselin, Polly Winsor, Jon Hodge, Gordon McOuat, David Johnson, Bernie Lightman, and last but not least Sharon Weltman for helping with the initial proofreading of the manuscript.

Chapter 1

A History of Nominalist Interpretation

Ever since the publication of Darwin's *Origin*, biologists, historians, and philosophers have interpreted Darwin as being a species nominalist. Species nominalism is the view that species are not real, that they are not out there in nature, existing irrespective of observation, but rather that they are man-made, like monetary currency or constellations, so that, from an objective, naturalistic point of view, they are real in name only.

This "received view" is based mainly on a literal reading of a number of passages in the *Origin*. In this chapter I shall begin by examining those passages. Following that I shall go back and examine Darwin's species concept(s) in his early period as an evolutionist, the period of his transmutation notebooks. I shall then proceed briefly up through the strata of his writings, trying to find where his supposed species nominalism began. I shall then take a brief excursion through the secondary literature, beginning with reviews of Darwin's *Origin* and proceeding right up to today. It will be interesting to see how the perception of Darwin as a species nominalist has been employed by a number of authors. Finally, I shall then examine how Darwin himself replied to the charge of species nominalism, as well as examine some other evidence which, together with what we shall see in subsequent chapters, should lead one to conclude that Darwin was in fact a species realist. In the very least, the end of this chapter along with the next four should bring to a close the easy days of finding in Darwin an ally for species nominalism.

Beginning with the *Origin* (1859),[1] in the concluding chapter Darwin proclaims that as a result of his investigations "we shall have to treat species in the same manner as those naturalists treat genera, who admit that genera are merely artificial combinations made for convenience. This may not be a cheering prospect; but we shall at least be freed from the vain search for the undiscovered and undiscoverable essence of the term species" (485). This passage relates to both halves of a modern distinction that partly defines the modern species problem, namely, the distinction between species as a taxon and species as a category, a distinction not always recognized but made much of by, for example, Ernst Mayr (e.g., 1987, 146). Again, briefly, species taxa are particular species, each of which is given a binomial, such as *Tyrannosaurus rex* or *Homo sapiens*. The species category, on the other hand, is the class of all species taxa. Among realists, the species category is captured in their respective definitions of the species concept. Thus, what is a genuine species according to one definition might not be counted as a genuine species according to another definition. A species nominalist, of course, would say that species definitions are ultimately arbitrary, because species taxa are ultimately arbitrary.

In the passage from Darwin's *Origin* quoted above, he seems quite clearly in the first part to assert that species taxa are unreal. He says that we shall have to treat species in the same way as genera nominalists treat genera, as not real but man-made, made simply for the sake of convenience.[2] In the second part of the passage, by referring to the "term species," Darwin seems clearly to be referring to the species category. There are other passages in the *Origin* that seem to second this view. For example, he says "From these remarks it will be seen that I look at the term species, as one arbitrarily given for the sake of convenience to a set of individuals closely resembling each other" (52). This passage is often quoted as supporting the interpretation of Darwin as a species nominalist, but it has to be remarked that the context of the passage makes it clear that Darwin is drawing his conclusion not from nature or from his own theory of evolution but from the taxonomic behavior of other naturalists. For in the previous paragraph he states that "If a variety were to flourish so as to exceed in numbers the parent species, it would then rank as the species, and the species as the variety" (52). This was not Darwin's view. Instead it was a practice common to his fellow naturalists.

Indeed, part of Darwin's overall argument for evolution was that in many cases expert naturalists could not themselves agree on whether a particular form was a variety or a species. For example, in the *Origin* he says "wherever many closely-allied species occur, there will be found many forms which some naturalists rank as distinct species, and some as varieties; these doubtful forms showing us the steps in the process of modification" (404; cf. 47, 49, 111, 248, 296–297). It was essential for Darwin that there be no clear distinction between species and varieties, otherwise varieties could not be what he called "incipient species" (52, 111, 114, 128), and the fact that expert naturalists could not agree in many cases on what is a species and what is a variety added a further prong in his attack on the fixity of species (in addition to his arguments from the fossil record, from biogeography, from embryology, from artificial breeding, etc.). And so again and again in the *Origin* we see Darwin assert that there is no essential or fundamental distinction between species and varieties. For example, he says "neither sterility nor fertility affords any clear distinction between species and varieties; but that the evidence from this source graduates away, and is doubtful in the same degree as is the evidence derived from other constitutional and structural differences" (248; cf. 51–52, 268, 272, 484–485).

A further part of Darwin's argument was that not only did naturalists in many cases disagree on what is a species and what is a variety, but they themselves could not agree on a definition of the species category. Even though, as Darwin early in the *Origin* recognized, "most naturalists" viewed species as "independently created" (6)—one might call this the common denominator[3]—they nevertheless gave "various definitions . . . of the term species" (44), definitions that concerned mainly the diagnostic criteria. This created a problem in itself, for as Darwin later in the *Origin* pointed out, "to discuss whether they ['many forms'] are rightly called species or varieties, before any definition of these terms has been generally accepted, is vainly to beat the air" (49).

And yet Darwin clearly recognized in the *Origin* the need for species talk. Consequently, on the issue of whether a particular form should be ranked as a species or a variety, he took the position that "the opinion of naturalists having sound judgment and wide experience seems the only guide to follow," and where they disagree the problem is to be settled simply by appealing to "a majority of naturalists" (47). This, of course, has an arbitrary ring to it. And indeed Darwin in the *Origin*, as we have seen earlier in this chapter, in apparent reference to his contemporaries, stated that he looks "at the term species, as one arbitrarily given for the sake of convenience" (52). Furthermore, again as we have seen earlier, in his concluding chapter Darwin took his own position that there is in fact no essential and fundamental distinction between species and varieties as a liberating one, since systematists "will not be incessantly haunted by the shadowy doubt whether this or that form be in essence a species" (484) and "shall at last be freed from the vain search for the undiscovered and undiscoverable essence of the term species" (485).

Small wonder, then, given all of the above, that scholars have commonly interpreted Darwin as a species nominalist, as we shall see later in this chapter. And yet how utterly odd, if those scholars are right, that Darwin would title his book *On the Origin of Species*, let alone with the addition *by Means of Natural Selection*! That the received view is wrong, that it is based on a superficial reading of Darwin, is something I shall argue later. For now, we need to ask when such apparently nominalist talk on Darwin's part began.

Certainly it did not start when Darwin began developing his evolutionary views. In his transmutation notebooks (Barrett *et al.* 1987) Darwin provides a number of definitions of "species," all realist in tone. Sometimes his definition is in terms of constant characters, as in Notebook B, begun in July 1837: "Definition of Species: one that remains at large with constant characters, together with other beings of very near structure" (213).

In other definitions Darwin focused on interbreeding, as in Notebook C, begun in March 1838 and finished in July of the same year: "A species is only fixed thing with reference to other living being—one species May have passed through a thousand changes, keeping distinct from other & if a first & last individual were put together, they would not according to all analogy breed together" (152). Darwin at some time later added an annotation to this page, writing "As species is real thing with regard to contemporaries—fertility must settle it." This page, both the original passage and the annotation, is interesting for its relation to the modern biological species concept made famous by Dobzhansky (1937) and especially Mayr (1942, 1970), which is based on interbreeding populations and genetic reproductive isolation mechanisms. What makes it interesting is not the emphasis on the fertility test. This was common in Darwin's time and before, having been made famous by Buffon (Lovejoy 1959). Instead, what is interesting about Darwin's passage is that as a species evolves radically over time, so that *vertically* in the Tree of Life it would in principle be incapable of interbreeding with its originals if they could be brought together, Darwin still insists on the reality of species at any given *horizontal* dimension, at any given cross-section in time. Like the modern biological species concept, Darwin, the evolutionist, provided a horizontal species concept and fixed the reality of species in the horizontal dimension, unlike a number of modern species concepts that insist on the vertical reality of species. What we shall see in Chapter 3 is that Darwin maintained this view in the *Origin* and that his main analogy for species (namely, languages) provides a powerful reason for believing that we today should also conceive of the reality of species primarily as horizontal rather than as vertical entities.

Interestingly, a little later in Notebook C Darwin seems to slightly change his mind, insisting now that "My definition of species. has nothing to do with hybridity,, is simply, an instinctive impulse to keep separate, which no doubt be overcome, but until it is the animals are distinct species" (161). However, earlier in Notebook B he had already used this criterion, when with regard to speciation he wrote that "repugnance to intermarriage—settles it" (24).

Indeed there can be little doubt that in his transmutation notebooks Darwin waffled between fertility/sterility and instinct. For example, in Notebook E, in an entry dated between October 16 and 19 of 1838, he states that "If they give up infertility in largest sense, as test of species.—they must deny species which is absurd.—their only escape is that rule applies to *wild* animals only. from which plain inference might be drawn that whole infertility was consequent on mind or instinct, now this is directly incorrect" (25). Similarly in his abstract of John Macculloch's *Proofs and Illustrations of the Attributes of God* (also in Barrett *et al.* 1987), which Darwin probably wrote in late 1838, he wrote "With respect to whether Galapagos beings are species, . . . it his highly unphilosophical to assert, that they are not species, until their breeding together has been tried" (167).

Except for his evolutionary perspective with his emphasis on the horizontal reality of species, Darwin in the above was doing nothing new. Indeed, years earlier James Prichard (1813, 3–15) provided a fairly detailed summary of the various criteria by which naturalists characterized species, which included not only constant character differences and the sterility test but also instinctual repugnance, immunological differences, and parasitological differences. It is not known whether Darwin had, by the time of the transmutation notebooks, read Prichard or a later edition, but it would not seem to matter, since he could have gotten the same ideas from other sources.

Turning now to Darwin's *Sketch of 1842* and his *Essay of 1844* (Darwin 1909), although they contain many ideas that are to be found later elaborated in the *Origin*, such as the idea that sterility is not an unfailing test, or that there are many forms about which expert naturalists cannot agree on whether they are species or varieties, there is not, unlike the *Origin*, any clear hint of species nominalism. Beginning with the *Sketch*, we find, actually, just like the transmutation notebooks, statements to the contrary. For example, Darwin says "Looking now to the affinities of organisms, without relation to their distribution, and taking all fossil and recent, we see the degrees of relationship are of different degrees and arbitrary—

sub-genera—genera—sub-families, families, orders and classes and king-doms" (35). Granted, Darwin is referring to his contemporaries, for he follows this passage with the sentence "The kind of classification which everyone feels is most correct is called the natural system, but no one can define this." Nevertheless, what is interesting is that with other higher taxa nominalists of his time (and most were higher taxa nominalists), Darwin does not include species in his list of arbitrary categories. Moreover, a little later in the *Sketch* he writes of "undoubted species" (48) and of "real species" (49). There is, however, immediately following, a passage that hints of species nominalism. In reference to "real species," which are distinct by every criterion, but admitting common descent, he writes "Can genera restrain us; many of the same arguments, which made us give up species, inexorably demand genera and families and orders to fall, and classes tottering. We ought to stop only when clear unity of type, independent of use and adaptation, ceases" (49). But here there is no reason to suppose that by the phrase "giving up species" Darwin means giving up the reality of species. A much more natural reading, given the basic presupposition of the majority of his antagonists, is "giving up the independent creation of species," or "giving up the fixity of species," which amounts to the same thing. Indeed we shall see in this and in later chapters that the phrase "giving up species" was a common one with Darwin, even though, again as we shall see in this and in later chapters, he did not give up their reality.

Turning now to the much longer *Essay*, beginning with the second chapter where he first uses the word "breed," Darwin added a note in the manuscript, writing "Here discuss *what a species is*" (81). However, unlike the *Origin*, Darwin did not follow through. Instead, much like the *Sketch*, even though he raises numerous problems for the independent creation of species and their fixity, and argues for evolution, he still continues to write of species as real. In fact, much like the *Sketch*, even though he denounces the categories of his contemporaries above the species level as "quite arbitrary" (202, 204–205), he continues to write of "true species" (204, 212, 241, 243, 246). And so unlike the *Origin,* with its many implied references to species as arbitrary, and with its concluding chapter which states that we shall have to treat species taxa as artificial and made for convenience, there is absolutely none of this in the *Essay*, neither in its nine argumentative chapters nor in its tenth concluding chapter.

Turning next to Darwin's correspondence, we do not find any clear signs of species nominalism until well into the 1850s. In fact, until that time, the impression we get from Darwin's correspondents is that most experienced naturalists believed in the reality of species, and Darwin, in turn, does not indicate that his view was otherwise. For example, in a letter from his closest friend and main correspondent, the botanist Joseph Dalton Hooker (September 4–9, 1845), Hooker wrote that "Those who have had most species pass under their hands as Bentham, Brown, Linneaus, Decaisne & Miquel, all I believe argue for the validity of *species* in nature" (Burkhardt and Smith 1987, 250). In his reply letter (September 10, 1845), Darwin recognized that "Lamarck is the only exception, that I can think of, of an accurate describer of species at least in the invertebrate kingdom, who has disbelieved in permanent species" (253). (As we shall see later in this chapter, Lamarck did more, and disbelieved in the reality of species.) Instead, the main effect that Hooker's letter had on Darwin was to doubt his own competence to theorize about species. Remarkably, Darwin wrote "How painfully (to me) true is your remark that no one has hardly a right to examine the question of species who has not minutely described many" (253). Darwin was especially taken by Hooker's extended criticism of the French writer Frédéric Gérard, who argued for species nominalism caused by his poor understanding of messy situations in nature and his lack of experience. (Indeed in an earlier letter to Darwin, written in late February 1845, Hooker states Gérard's species nominalism explicitly, and offers to send Darwin a copy of Gérard's tract; Burkhardt and Smith 1987, 149.) And even though Hooker would reply that he by no means meant to imply that Darwin was not in a position to theorize about species, Darwin did not attempt to procure species nominalism from his evolutionary theories. Instead, toward the end of 1846 he embarked on an eight-year taxonomic study of barnacles, which he completed in September of 1854.

Much of what is interesting about Darwin's work on barnacles is that he was struck by the variability of organisms. What we shall see in chapter 3, when we focus on his published works on barnacles, is that even though he stressed that variability, he did not talk of species as if they were arbitrary. Instead he argued that most species of barnacles, even when minutely studied, turn out to be taxonomically good species. Equally revealing is what Darwin wrote in his correspondence. However problematic was the variability of barnacles taxonomically, Darwin still did not espouse species nominalism. The following reply letter to Hooker (June 13, 1850) perfectly captures Darwin's thinking throughout this period:

> You ask what effect studying species has had on my variation theories; I do not think much; I have felt some difficulties more; on the other hand I have been struck (& probably unfairly from the class) with the variability of every part in some slight degree of every species: when the same

organ is *rigorously* compared in many individuals I always find some slight
variability, & consequently that the diagnosis of species from minute differences is always dangerous. I had thought the same parts, of the same
species more resembled than they do anyhow in Cirripedia, objects cast
in the same mould. Systematic work w^d be easy were it not for this confounded variation, which, however, is pleasant to me as a speculatist
though odious to me as a systematist. [Burkhardt and Smith 1988, 344]

Other letters confirm this view. For example, earlier in the same year, in a
letter to J.J. Steenstrup (January 25, 1850), Darwin wrote "I much dislike
giving specific names to *each* separate valve, & thereby almost certainly
making three or four *nominal* species for each true species" (Burkhardt and
Smith 1988, 306).

   Granted, toward the end of his work on barnacles, Darwin had become quite tired of detailed species work, so much so that he started to
sound like he might be swaying to species nominalism. In an often-quoted
letter to Hooker (September 25, 1853), Darwin wrote

In my own cirripedial work (by the way, thank you for the dose of soft
solder [i.e., flattery—*OED*], it does one, (or at least me) a great deal of
good,—in my own work, I have not felt conscious that disbelieving in the
permanence of species has made much difference one way or the other; in
some few cases (if publishing avowedly on doctrine of non-permanence)
I sh^d. *not* have affixed names, & in some few cases sh^d. have affixed names
to remarkable varieties. Certainly I have felt it humiliating, discussing &
doubting & examining over & over again, when in my own mind, the
only doubt has been, whether the form varied *today or yesterday* (to put a
fine point on it, as Snagsby would say). After describing a set of forms,
as distinct species, tearing up my M.S., & making them one species; tearing that up and making them separate, & then making them one again
(which has happened to me) I have gnashed my teeth, cursed species, &
asked what sin I had committed to be so punished: But I must confess,
that perhaps nearly the same thing w^d. have happened to me on any
scheme of work. [Burkhardt and Smith 1989, 155–156]

What is typically overlooked, however, is what Darwin says to Hooker at the
very end of his letter: "whether you make the species hold up their heads or
hang them down, as long as you don't quite annihilate them or make them
quite permanent; it will all be nuts to me [i.e., a source of pleasure or delight—
OED]." Darwin was not yet talking the language of species nominalism.
   The fact is, we don't first start to find species nominalism talk in Darwin until we turn to his long though unfinished book on species evolution,

titled *Natural Selection*, which was begun in mid-1856 and stopped on June 18, 1858, when Darwin received the letter from Alfred Russel Wallace basically anticipating Darwin's views (which was of course to spark the writing of his *Abstract*, later to be titled *On the Origin of Species*). As Stauffer (1975, 7–9) points out, Darwin waited roughly 20 years to publish his evolutionary views because he wanted to present a strong *scientific* case for evolution (more particularly, evolution by natural selection and divergence) and thus avoid the scientific ridicule heaped upon earlier writers on evolution, in particular, Lamarck and Chambers. Before we turn to *Natural Selection*, though, we have to wonder why Darwin would wait so long to start sounding like a species nominalist.

One theory that might suggest itself follows from the important work of Dov Ospovat (though I know of no one who has used Ospovat to develop it). According to Ospovat (1981), from the time Darwin hit upon natural selection in his transmutation notebooks, through the *Sketch* and *Essay*, and until he had finished his barnacle work, Darwin shared with his contemporaries the belief (which was theologically based) in harmony and perfect adaptation in nature, with variation being minor, so that in his view natural selection worked only intermittently, in those periods when changes in conditions meant that a species was no longer perfectly adapted to its environment. Between September 1854, however, and June 1858, when he received the shocking letter from Wallace, Darwin's view on variation and adaptation gradually though radically changed, from perfect adaptation with intermittent natural selection to imperfect adaptation with continuous natural selection. One might think that this new view would have occurred to him early on in his barnacle work. But Ospovat (ch. 7) argues that it was not until after Darwin finished his barnacle work that he sat down to seriously rethink his theory of evolution. The main problem was to explain the treelike, group nested in group, hierarchical classification schemes of his fellow naturalists. Darwin's solution was his principle of divergence, which he developed in the period from 1854 to 1858 and which, in a letter to Hooker (June 8, 1858), he called (along with natural selection) "the key-stone of my Book" (Burkhardt and Smith 1991, 102). According to this principle, wide-ranging species will typically be exposed within their range to a variety of conditions, most importantly to empty niches (to use modern terminology) which they will tend to fill, and hence evolve in a branchlike fashion.

Based on Ospovat's work, then, one might conjecture that prior to 1854—prior to when Darwin started rethinking his theory and still believed in perfect adaptation with only intermittent natural selection—Darwin would naturally think that species are real so long as natural selection is not working upon them (possibly there was an influence from Lamarck here; cf. note 6), so that once his view changed to imperfect adaptation with continuous natural selection he consequently became a species nominalist.

This is an interesting conjecture. Unfortunately it fails for the fact, as we shall see in this and in later chapters, that under his skin Darwin was not really a species nominalist, not in his post-barnacle period nor in the *Origin* or beyond.

There is something else, however, which went on in the period between 1854 and 1858, which does help to fully explain Darwin's species nominalism talk, begun in *Natural Selection*. The evidence is in his correspondence. As pointed out earlier, Darwin began his big species book, what was to be his heavily detailed case for evolution to the world, in mid-1856, and the problem was to avoid the scientific ridicule heaped upon the mainly speculative attempts of earlier writers, in particular Lamarck and Chambers. The problem, in short, was to convince expert naturalists more than anyone else. What Darwin got from botanist correspondents such as Hooker, but mainly from Hewett Cottrell Watson, was that the very concept of species itself was a major impediment to convincing the scientific world that species are not fixed but evolve.

Interesting in this regard is a letter from Hooker (July 8, 1855), in which Hooker comments on a Himalayan thistle intermediate between two common species of English thistles. Hooker writes "The more I study the more vague my conception of a species grows, & I have given up caring whether they are all pups of one generic type or not" (Burkhardt and Smith 1989, 372). Hooker goes on to say that not caring anymore whether this or that is a real species forms no impediment to tracing character distribution and discovering the laws of distribution, which he thinks "is certainly all we can expect to prove in our day" (372). Here Darwin may have begun to realize that the species concept, when trying to get his evolutionary views across, presented more of an impediment than anything else, and so was best bypassed. And yet, interestingly, when we turn to chapter 8, we shall see that in his correspondence, when Darwin is trying to convince an important naturalist of his views on evolution, he uses the language of species nominalism, but only until he is convinced that he had a convert, after which time he reverted to the language of species realism. (Indeed, as I argue later in this chapter and in chapters 2 through 6, Darwin had an implicit but fairly clear species concept that was both realist and evolutionary.)

If Hooker's letter did not make Darwin think of the value of not getting bogged down on the topic of what a species really is when presenting his scientific case to the world for evolution, one of the letters from Watson almost certainly did. Watson, who was converted to evolution (Watson 1845a) shortly after reading Chambers' *Vestiges*, anonymously published in 1844, wrote to Darwin (August 13, 1855) that "The grand difficulty for naturalists or botanists of our turn of thought, is, that the use of the word '*species*' by technical describers is indefinite & variable. . . . Practically, it means only an idea of the mind, with no more real restriction in its application to objects, than have the words '*genus*' or '*order*.'" Watson then cites Hooker and the French botanist Alexis Jordan as examples of lumpers and splitters respectively (the former grouping varieties into species, the latter making a species out of the smallest variety). Watson goes on to say "In all my attempts to advance geographical botany, I am stopt by the application & signification of the word 'Species.' Where I seek to effect precise comparisons of objects & numbers & proportions,—that word constantly frustrates & makes vague & indefinite" (Burkhardt and Smith 1989, 406).

Indeed, turning now to *Natural Selection*, we can see the influence of Watson, on both the "grand difficulty" presented by the variability of species concepts in Darwin's contemporaries, as well as the implicit suggestion that it is better to bypass the concept altogether.[4] First, in a choice of words echoed shortly after in the *Origin*, he says "In the following pages I mean by species, those collections of individuals, which have commonly been so designated by naturalists" (Stauffer 1975, 98).

What is equally interesting is what Darwin wrote immediately before this:

> . . . how various are the ideas, that enter into the minds of naturalists when speaking of species. With some, resemblance is the reigning idea & descent goes for little; with others descent is the infallible criterion; with others resemblance goes for almost nothing, & Creation is everything; with others sterility in crossed forms is an unfailing test, whilst with others it is regarded of no value. At the end of this chapter, it will be seen that according to the views, which we have to discuss in this volume, it is no wonder that there should be difficulty in defining the difference between a species & a variety;—there being no essential, only an arbitrary difference. [Stauffer 1975, 98]

This passage compares, interestingly, with a letter Darwin wrote to Hooker (December 24, 1856) at roughly the same time:

> I have just been comparing definitions of species, & stating briefly how systematic naturalists work out their subject: . . . It is really laughable to see what different ideas are prominent in various naturalists minds, when they speak of "species" in some resemblance is everything & descent of

little weight—in some resemblance seems to go for nothing & Creation the reigning idea—in some descent the key—in some sterility an unfailing test, with others not worth a farthing. It all comes, I believe, from trying to define the undefinable. [Burkhardt and Smith 1990, 309]

In later chapters, after examining what I believe to be Darwin's objective set of criteria for delimiting species taxa, only then will the disingenuous nature of these passages become apparent, especially when put in their context, and only then will it make sense to develop in detail a strategy theory to explain them (chapter 8).

For the present, it will be useful to examine how reviewers of the *Origin* responded to the apparent species nominalism of that book. The first point to notice, using late 1859 and 1860 as typical, is that many if not most of the reviewers simply bypassed the issue of Darwin's apparent species nominalism. They didn't so much as even mention it. Instead they focused on Darwin's argument for evolution, in the main rejecting it (e.g., Anon. 1859; Crawfurd 1859; Leifchild 1859; Murray 1859; Anon. 1860a; Anon. 1860b; Bowen 1860; Haughton 1860; Sedgwick 1860; Simpson 1860; Wilberforce 1860).

Even among Darwin's supporters, his apparent species nominalism was typically ignored (e.g., Chambers 1859; Hooker 1859; Huxley 1859b, 1859c, 1860a, 1860b; Carpenter 1860; Gray 1860b).

Returning to his critics, there were some, however, who did indeed take Darwin's apparent species nominalism to be in fact his position. For example, Louis Agassiz (1860b) raised what seemed to him a perfectly logical point: "If species do not exist at all, as the supporters of the transmutation theory maintain, how can they vary? and if individuals alone exist, how can the differences which may be observed among them prove the variability of species?" (143). Richard Owen (1860) claimed that "on the hypothesis of 'natural selection' . . . . the species, like every other group, is a mere creature of the brain; it is no longer from nature" (532), which he rejects on what he calls "present evidence from form, structure, and procreative phenomena." Instead he agrees with the Linnean axiom that species are the work of nature, which he quotes as "*Classis et Ordo* est sapientiæ, *Species* naturæ opus" (532). Thomas Vernon Wollaston (1860) referred explicitly to a page of the *Origin* where we find apparent species nominalism and wrote that "it is no sign of metaphysical clearness when our author (p. 51) refuses to acknowledge any kind of difference between 'genera,' 'species,' and 'varieties,' except one of *degree*" (133), which was, he continued, "to throw doubt on a distinction between essentially different *ideas*"

(134). Others, without claiming or implying that their interpretation of Darwin's species nominalism came explicitly from Darwin himself in the *Origin*, claimed that species nominalism followed from his theory of evolution. John Dawson (1860), for example, claimed that Darwin's book "seeks . . . to reduce all species to mere varieties of ancient and perhaps perished prototypes" (101) and that with his doctrine we "break down the distinction between species and varieties as to deprive our classifications of any real value" (119). Similarly, William Hopkins (1860) wrote that "all theories—like those of Lamarck and Mr. Darwin—which assert the derivation of all classes of animals from one origin, do, in fact, deny the existence of natural species at all," where by "natural species" he means "the grouping is formed by nature," whereas with "artificial species" the grouping is "arbitrary" (747).

Among Darwin's supporters, so too did some recognize his species nominalism, although often they did not actually quote Darwin as such but inferred it from his views. Asa Gray (1860a), for example, argued that it follows from Darwin's theory that whether the human races constitute one species or more is to be settled "according to the notions of each naturalist as to what differences are specific" (158). Interestingly, against Agassiz, who in his species concept "discards the idea of a common descent as the real bond of union among the individuals of a species, and also the idea of a local origin,—supposing, instead, that each species originated simultaneously, generally speaking over the whole geographical area it now occupies or has occupied" (155), Gray claims that his (Agassiz's) theory equally makes species "subjective and ideal" (158)! This is an interesting use of Darwin's species nominalism. Henry Fawcett (1860) too, although he did not quote anything from the *Origin* as espousing species nominalism, implied that it also followed from Darwin's view. Repeating (though magnifying) the radical disagreement between Babington and Bentham on the number of species of English plants (cf. *Origin*, 48), Fawcett writes "The question of species may thus, at the first sight, appear to be a dispute about an arbitrary classification, and it may naturally be asked, Why, therefore, does the problem of the Origin of Species assume an aspect of supreme scientific interest?" (82). Similarly George Henry Lewes (1860), likewise feeding off the disagreement between naturalists over whether a particular form is a species or a variety (which of course Darwin himself made much of in the *Origin*), writes that "The reason of this uncertainty is that the *thing* Species does not exist: the term expresses an *abstraction*, like Virtue or Whiteness; not a definite concrete reality, which can be separated from other things, and always be found the same" (443).[5]

What we have to keep in mind in all of this is that in Darwin's time, so unlike today, the equation of evolution with species nominalism was deeply entrenched. And arguably it was Lamarck who began this equation. In the first chapter of his book on evolution (Lamarck 1809), he states that all divisions of nature into classes, orders, families, genera, and species are "artificial devices" (20), that "they appear to derive from certain apparently isolated portions of the natural series with which we are acquainted," and that nature has produced "only individuals who succeed one another and resemble those from which they sprung." The relation of individual organisms to the natural series, he immediately goes on to say, is that "these individuals belong to infinitely diversified races; which blend together every variety of form and degree of organisation; and this is maintained by each without variation, so long as no cause of change acts upon them" (21).[6]

Consequently we find Charles Lyell (1832), as he begins his long critique of Lamarck's evolutionism, state the issue as "whether species have a real and permanent existence in nature; or whether they are capable . . . of being indefinitely modified in the course of a long series of gradations?" (1; cf. 23), which, following his critique, he concludes that "it appears that species have a real existence in nature, and that each was endowed, at the time of its creation, with the attributes and organization by which it is now distinguished" (65). This dichotomy—either species are permanent and therefore real or impermanent and therefore unreal—is repeated again and again in the literature of Darwin's time. For example, William Whewell (1837 III) wrote that "in short, *species have a real existence in nature*, and a transmutation from one to another does not exist" (576). A further example is Watson (1845b), who after arguing empirically about the mutability of primroses and cowslips, wrote that "If we allow the cowslip and primrose to be two species, and yet allow that one can pass into the other, either directly or through the intermediate oxlip, we abandon the definition of species, as usually given, and fall into the transition-of-species theory. . . . Let a few other cases be adduced, between reputed species equally similar, and we shall be forced to recast our ideas and definition of the term 'species.' It would unavoidably become arbitrary and conventional; with no more exactness or constancy of application, than we can give to the terms 'genus' or 'order'" (219). As one final example, Wollaston (1860) claimed that either species are permanent and real (the traditional species concept) or else we are left with "the otherwise hopeless task of understanding what a species really is" (133), which may be taken as an epistemological assertion only, but possibly also as an ontological one.

So it was easy and natural for reviewers to read species nominalism in Darwin's *Origin* and to see no need to scratch beneath the surface. In later commentators on Darwin's *Origin*, however, living in a different scientific milieu, what we often find is that those who interpret Darwin as a species nominalist do so to use Darwin as an imprimatur for their own nominalist arguments. We shall also find, of course, that they just plain overlooked the evidence for Darwin's species realism.

A good example to begin with is E.B. Poulton (1903), a naturalist and selectionist whom Mayr (1982, 272) took to be a "pioneer" of the biological species concept. In reply to Max Müller, who claimed that in spite of the title of the *Origin* Darwin never gave us a species concept, Poulton (78) replies that Darwin did and that it is given at the end of the *Origin* where he says "Systematists will have only to decide (not that this will be easy) whether any form be sufficiently constant and distinct from other forms, to be capable of definition; and if definable, whether the differences be sufficiently important to deserve a specific name" (484). Throughout his paper Poulton gives the impression that Darwin was a precursor of the syngamic species concept which he himself prefers, "syngamic" meaning "free interbreeding under natural conditions" (90), an "inter-breeding community" (94). But Poulton does nothing to elaborate on Darwin's species concept, for example, whether it includes sterility between forms or even whether indeed Darwin himself thought that species are fully syngamic. Instead, he repeatedly emphasizes the "subjective character" (89), the "subjective element" (92), the "subjective criterion" (93) in Darwin's species concept. This, however, is not to attribute to Darwin species realism. Indeed naturalists at this time tended to read Darwin as a species nominalist (e.g., Arthur 1908, 244, who quotes Darwin approvingly; Cowles 1908, 267), in conformity with the species nominalism of the time (e.g., Morgan 1903, 33; Bessey 1908, 218; Coulter 1908, 272).

On the other side of the coin, keeping to the pre-Modern Synthesis era, we have the geneticists, who were principally saltationists and tended to be species nominalists (Mayr 1957a, 4–5, 1982, 540–550). I have found it impossible, however, to find any of them quote Darwin as a species nominalist, which makes sense since they were anti-selectionists and so therefore would be unlikely to appeal to Darwin as an authority on the matter.

Turning now to the post-Synthesis period, it is remarkable to find biologists, philosophers, and historians repeatedly ascribe to Darwin species nominalism. A good example to begin with is the geneticist J.B.S. Haldane, together with Fisher and Wright one of the three main founders of the Modern Synthesis. In his contribution to a symposium on the species concept in paleontology, Haldane (1956) states at the outset that "I share the views of Darwin" (95), which he goes on to elaborate as being that "A species . . . is a name given to a group of organisms for convenience, and indeed of necessity" (95), and moreover that "the concept of a species is a concession to our linguistic habits and neurological mechanisms" (96). Seeing species in both space and time, he adds that "in a complete pale-ontology all taxonomic distinctions would be as arbitrary as the division of a road by milestones" (96). As we shall see in subsequent chapters, how-ever, this view fails to recognize that Darwin thought of species as primar-ily horizontal entities and as being delimited in the main by natural selection, which is a far cry from the subjectivity that Haldane ascribes to Darwin's view.

In many ways a more important example is the ornithologist Ernst Mayr (1957a), according to whom "In Darwin, as the idea of evolution became firmly fixed in his mind, so grew his conviction that this should make it impossible to delimit species. He finally regarded species as some-thing purely arbitrary and subjective" (4; cf. Grant 1957, 58–59, for the same view expressed in the same volume, and also Mayr 1970, 13, 1976, 259, 1991, 30). What is interesting about Mayr is not that he was using Darwin as the imprimatur for his own view (Mayr was, after all, a hardcore species realist), but that he would later blame Darwin's species nominalism on Darwin's association with botanists. In explanation of Darwin's mature view of species as "purely arbitrary designations" (269), as opposed to Dar-win's earlier view in the 1830s which "was very close to the modern biolog-ical species concept" (266), Mayr wrote that "His reading as well as his correspondence indicate that after 1840, and particularly from the 1850s on, Darwin was increasingly influenced by the botanical literature" (267), and he goes on to quote William Herbert (a leading English authority on plant hybridization), for whom he says "the genus was the only 'natural' cat-egory" and of whom he says "Perhaps no other botanist influenced Dar-win's thinking more" (268).

There are at least two problems with this view, however. The first one concerns Herbert in particular. Darwin had indeed read Herbert (his *Amaryllidaceæ* is frequently cited in Darwin's Notebook E), had exchanged a number of letters with him in mid-1839, and had even visited him once in September 1845 (Herbert died in 1847). Equally important, in the *Ori-gin* Darwin favorably refers to Herbert on the topic of the struggle for ex-istence among plants (62), and even more favorably on the topic of perfect fertility in interspecific hybrids in the genera *Crinum* and *Hippeastrum*

(249–251). For Herbert, hybrids are sometimes very fertile, so that the distinction between species and varieties has "no real or natural line of difference" (Burkhardt and Smith 1986, xvii–xviii, 182 n. 1). His species nominalism, however, if indeed it was such, followed apparently from taking sterility as the defining criterion of species. It was like Lyell, who thought that if evolution is true then species must be unreal. But as we shall see, in spite of Darwin's acceptance of evolution and of the non-universality of the sterility of hybrids, he nevertheless thought that species were real (a view shared, of course, with most biologists today). Moreover, in his correspondence Darwin does not seem particularly impressed by Herbert's expertise. For example, in a letter to Hooker (October 28, 1845) written shortly after visiting Herbert, Darwin remarks that Herbert "knows surprisingly little what others have done on same subjects" (Burkhardt and Smith 1987, 261).

But even more importantly against Mayr, Darwin repeatedly tells us that most of his contemporary naturalists were species realists. For example, near the beginning of the *Origin* Darwin tells us that "the view which most naturalists entertain" is that "each species has been independently created" (6). Later in the *Origin* he gives specific names in the fields of paleontology and geology, stating that "all the most eminent paleontologists" and "all of our greatest geologists . . . have unanimously, often vehemently, maintained the immutability of species" (310). But we should not take this to mean that Darwin did not think the same was true of botanists. As we have seen earlier in this chapter, Bentham, Hooker, Gray, and even Watson (each of them eminent botanists, with the latter three being Darwin's main botanist correspondents) were species realists. Moreover, that the vast majority of eminent botanists were species realists had been driven home to Darwin a number of times. For example, Watson (1843) states that there is a consensus among British botanists that although "genera are allowed to be purely conventional groups, . . . species are commonly believed to have a distinct and permanent existence in nature" (613). Moreover there is Hooker's letter to Darwin (September 4–9, 1845) which we have seen earlier, in which he wrote "Those who have had most species pass under their hands as Bentham, Brown, Linnaeus, Decaisne & Miquel, all I believe argue for the validity of *species* in nature" (Burkhardt and Smith 1987, 250). Each member of this list was a first-rate botanist. In sum, all of this adds credence to Darwin's remark in his autobiography (1876a), when looking back at his pre-*Origin* days, that "I occasionally sounded not a few naturalists, and never happened to come across a single one who seemed to doubt about the permanence of species. Even Lyell and Hooker, . . ." (124).

Given the above evidence, it is quite possible that Mayr, then, in blaming the influence of Darwin's botanist correspondents, was actually projecting onto history his own problems with botanists, for many modern botanists have argued that the biological species concept (endorsed more strongly by Mayr than by anyone else) applies poorly to the world of plants, a claim that Mayr was long eager to discount and that he attempted to refute by studying a local flora (Mayr 1992). In chapter 6, I shall examine the views of some of these modern botanist critics of a reproductive criterion for species.

What is interesting for our purposes here is that one of them, Donald Levin (1979), in arguing that the biological species concept does not apply well to plants, argues consequently for species nominalism—"plant species are utilitarian mental constructs" (381)—and quotes Darwin in support. As he puts it, "Darwin concurs with Locke" (382; cf. Cowan 1962, 434–435, for the same equation). John Locke, of course, is famous for arguing in his *Essay Concerning Human Understanding*, first published in 1689, that our species designations are not made by nature but by ourselves, that species words simply refer to our abstract ideas produced by abstracting what is common from a number of individuals. Thus, for Locke, "this is a Man, that a Drill [baboon]: And in this, I think, consists the whole business of *Genus* and *Species*" (cf. Stamos 2003, 40–47). In an earlier work (Stamos 1996, 128–129), in reply to Antony Flew who believed that Darwin never read Locke, I not only cited a source to the contrary, but quoted an interesting passage from Darwin's Notebook M (84), in which he wrote, "Origin of man now proved.—Metaphysic must flourish.—He who understands baboon would do more towards metaphysics than Locke" (Barrett *et al.* 1987, 285). Although Notebook M was devoted to the metaphysics of mind, it is quite possible that in this passage Darwin was referring to Locke's species nominalism as well as to his own rejection of that view. What we have to keep in mind is that Darwin in his transmutation notebooks, as we have seen, was a species realist. What we shall see in subsequent chapters is that he never, not even in his mature period, concurred with Locke.

In the above we have looked at three biologists who read Darwin as a species nominalist. There are, of course, many more (e.g., Gould 1980, 205–206; Wiley 1981, 41; Howard 1982, 17, 37; Rieppel 1986, 304, 307; Eldredge 1989, 109–110; Luckow 1995, 590). And among philosophers the same view naturally persists. For example, the philosopher Elliott Sober (1993) wrote that Darwin's book should have been titled "*On the Unreality of Species as Shown by Natural Selection*" (143). (For other examples of philosophers who share this view, cf. Hull 1965, 203; Thompson 1989, 8; Ereshefsky 1992a, 190).

Historians are interesting here in a slightly different way. Alvar Ellegård (1958, 200), for example, repeats the same view. In fact, the winds of change did not begin to blow until the biologist Michael Ghiselin (1969) argued that for Darwin species taxa are real but not the species category, that Darwin was in one sense a species realist but in another sense a nominalist, so that Darwin did not have a species concept/definition. A number of years later the philosopher John Beatty (1985), following a suggestion by Frank Sulloway (1979), added a strategy theory to Ghiselin's thesis to explain why Darwin in the *Origin* would repeatedly define species nominalistically and yet in fact hold that species taxa are real. Historians have seemed to simply follow this lead. Jon Hodge (1987), for example, as well as Gordon McOuatt (1996, 2001), both subscribe to the Ghiselin/Beatty thesis, while attempting to provide their own twists. I shall return to these authors in chapter 8, where I develop my own strategy theory. What is interesting to note at this point is that among professional historians, and increasingly among philosophers (e.g., Kitcher 1993, 32 n. 45; Laporte 2004, 192 n. 13; Grene and Depew 2004, 213), the Ghiselin/Beatty thesis has become the received view (cf. chapter 8).[7]

What I shall attempt to do in the following chapters is to take the now received view—that Darwin was a species taxa realist but not a species category realist—to the next level, that is, to show that he was in fact a species category realist, that when he looked at taxa he had an implicit species concept that he applied again and again. But that is not all that I shall do.

Before we begin, however, it is important to finish off this chapter with some strong evidence, direct and indirect, that Darwin's view in the *Origin* and beyond was not that of a species nominalist, in other words that he did not think of species as akin to constellations, the standard example of nominalism (e.g., Lyell 1832, 19; Darwin 1859, 411), where the individuals are real but the groupings of them are subjective and arbitrary. A good place to begin is with Darwin's reply to Agassiz's quip that if species are not real then it makes no sense to say they vary. In a letter to Asa Gray (August 11, 1860) Darwin wrote "I am surprised that Agassiz did not succeed in writing something better. How absurd that logical quibble;—'if species do not exist how can they vary?' As if anyone doubted their *temporary* existence" (Burkhardt *et al.* 1993, 317, italics mine). Moreover, in the margin of his copy of Agassiz's review, where Agassiz's quip is to be found, Darwin wrote "exist only temporarily" (Burkhardt *et al.* 1993, 318 n. 4). Temporary existence is, of course, nonetheless real existence, not nominal existence, and in chapter 3 I shall attempt to determine exactly what Darwin meant.[8]

There is other evidence as well. In the *Origin* itself, Darwin wrote "I believe that species come to be tolerably well-defined objects, and do not at any one period present an inextricable chaos of varying and intermediate links" (177). This passage, along with many others, will help to establish an important part of what Darwin thought on the ontology of species, again as well shall see in chapter 3. Another piece of good evidence is to be found in Darwin's letter to Hooker (October 22, 1864), in which he wrote "The power of remaining for a good long period constant, I look at as the essence of a species, combined with an appreciable amount of difference; & no one can say there is not this amount of difference between Primrose & Cowslip" (Burkhardt *et al.* 2001, 376). Without the last clause, this passage has the power to mislead, as, for example, it did Poulton (1903, 91). Both parts together, however, help to determine a further important feature of Darwin's mature species concept, as we shall see in Chapter 5. As a final piece of evidence that should suffice for the present, Darwin in one of his articles (1863b) calls it a "great truth," regardless of evolutionary mechanisms, "that species have descended from other species and have not been created immutable" (81).

It remains now to determine exactly—inasmuch as that is possible—what Darwin meant when he wrote of "species."

Chapter 2

Taxon, Category, and Laws of Nature

As pointed out earlier, in the literature on the modern species problem arguably the most fundamental distinction is between species as a taxon and species as a category. Species taxa are particular species and are given binomial names such as *Tyrannosaurus rex* and *Homo sapiens*. It is species taxa that evolve, that speciate, that have ranges, that are broad-niched or narrow-niched, and that become endangered or extinct. The species category, on the other hand, is the class of all species taxa.

The species problem is typically phrased in terms of determining the ontology of species taxa. But the species category is also a problem. Some conceive of it as an abstract class, a class captured by a definition, the definition determining (or capturing) membership in the class, such that the class (along with its definition) stays the same through time as particular species taxa come and go in terms of existence. Others, however, conceive of the species category not as an abstract class but as a concrete set, simply the set of all species taxa (for examples of both views, cf. Stamos 2003, 150, 258).

This latter view, however, that the species category is a set rather than a class, has serious problems typically overlooked by its advocates. Granted, it has greater parsimony, since the species category *simply is* its member taxa and nothing more. But this view comes at a great cost. For one thing, it cannot possibly capture the sense of the species category remaining the same while its member taxa come and go. If the species category is simply a set, then as a set it changes every time a species taxon comes into existence or goes out of existence. If the species category is viewed as an abstract class, on the other hand, it does not have this problem. Just as with the category gold, which stays the same even though particular atoms of gold come and go in terms of existence, so too, on this view, the species category stays the same even though particular species taxa come and go in terms of existence.

But there is a further problem for the view that the species category is a set, namely, that it precludes laws of nature for the species category. It has become a staple of modern philosophy of biology that there are no laws of nature for particular species taxa. There are no laws, for example, specifically for *Tyrannosaurus rex*. And this absence is exactly what one finds in modern biological literature. Moreover, this observation makes sense from a number of key points. Specifically, a species taxon cannot be the subject of a law of nature because a species taxon gradually evolves. Laws, however, on the usual view of laws, do not change or evolve. The law of gravity, arguably, has not been changing over time, and the same holds true for the speed of light in a vacuum (cf. Nagel 1961, 378–380; Armstrong 1983; Weinert 1995; Stamos 2003, 215–220). But is the species *category* the subject of laws of nature? This is indeed a very live issue in modern biology. The problem is that if the species category is conceived to be a set, then debate over whether the species category is the subject of laws of nature is immediately shut off. If the species category is a set, then it cannot possibly be the subject of laws of nature because the species category, as a set, is gradually changing over time, whereas laws of nature do not change over time. If the species category is conceived to be an abstract class, on the other hand, then it could possibly be the subject of laws of nature. It need not be the subject of laws of nature, but the possibility is not precluded, given the common view of laws as unchanging.

Perhaps the main attraction of conceiving of categories as sets, as nothing more than their members, is that it conforms with a materialist philosophy of nature, according to which what is real is, or is fundamentally, material. My impression is that many scientists (and not just scientists) are typically too materialistic to take seriously the view that abstract categories are part of the fabric of nature. But this should not be considered such a far-out view (cf. Weinberg 1992, 46). In cosmology, it is debated among physicists whether the so-called cosmic constants (e.g., the law of gravity, or the strong nuclear force, which holds protons and neutrons together in the nucleus of an atom) are truly universal or are different in different pockets of the universe (sometimes called "baby universes"). One idea is that the Big Bang could have produced a different set of cosmic constants. The set that it did produce obviously allowed for the formation of stars and planets, with life eventually evolving on at least one of the planets. This view can easily be interpreted in terms of abstract categories built into the fabric of the universe. The Big Bang, as it actually happened some 15 billion years ago, resulted in the categories of the chemical elements, each element being a category. Many of these elements, such as gold, have naturally occurring physical members, namely, each and every atom that has 79 protons in its nucleus. Some of the other categories of chemical elements, on the other hand, have no naturally occurring physical members in the universe. They are empty categories, their membership being nothing more than a permanent possibility. With some of these categories, of course, humans in recent years have supplied membership, the highest so far being (I believe) element 114. But humans did not create this chemical element/category. The Big Bang did. Humans, instead, created an atom that simply fit the category (even if it was only for a split second). Presumably, because of the strong nuclear force and the mass of protons and neutrons, there is a physical limit as to how many protons a nucleus can have, even for a split second. Whatever that limit is, it exists, as an abstract reality, out there in nature, even though humans have not yet discovered it.

Similarly one can think of the species category, as an abstract permanent possibility of nature, coming into existence at the time of the Big Bang. Richard Dawkins (1983) alludes to this idea when he refers to what he calls "Universal Darwinism," the idea that natural selection must be the main mode of species change wherever in the universe there are species.

Had the Big Bang happened differently, had it produced a different set of cosmic constants, it is quite possible, even probable, that the species category would not exist, that it would not be part of the fabric of the universe, its having members not a possibility at all. This would occur if the set of cosmic constants resulting from the Big Bang would not allow for the formation of stars. Without the formation of stars, there could not be planets. And without planets, life could not evolve and species would be impossible.

Fortunately that is not the universe we have. It is not only a physical universe, but it also contains abstract categories built into it. Many of these categories are the chemical elements. The chemical element categories are strictly essentialistic. Any atom that has 79 protons in its nucleus, for example, is an atom of gold, while any atom that does not is not an atom of gold. Another category is the species category.

The idea that categories are objective, out there in nature, irrespective of and independent of minds, is of course a controversial one in the history of thought. There are basically two traditions. Kant thought of categories as existing only in the mind. On the nature of nature in itself, the noumenal world, he was agnostic. Human perception is conditioned by the categories of the mind, so that reason can go no further than the phenomenal world, which is itself the joint product of the mind and nature in itself. This kind of view, however, is alien to the modern scientific mind, which is typically realist in orientation. Scientists discover the way things really are, not simply the ways in which the mind creates the phenomenal world. They discover real laws, laws in nature, as well as processes and histories and entities and kinds in nature. In this sense modern scientists are much more in line with Aristotle. On the topic of categories, Aristotle thought of them as ways of being, and he specifically listed primary substances (concrete individuals) and secondary substances (*eidos* and *genos*) among them, examples of the latter being man, horse, and animal (*Categories* 1^{b}25–2^{a}19). Moreover, he explicitly claimed that of all the ways of being (such as property, relation, quantity, having, and doing) only substances can exist independently (*Physics* 185^{a}31–32). (For more on Aristotle on species, cf. Stamos 2003, 102–113; Grene and Depew 2004, 1–34.)

This idea, that there are objective categories in nature, in particular for species and for other categories, was continued by Carolus Linnaeus. This self-proclaimed Prince of Botanists, who received his education in Swedish universities, which were themselves among the last strongholds of Aristotelianism in Europe, thought of categories as boxes. As Leikola (1987) puts it,

> His great idea was a rigid and homogeneous categorizing: everything in nature should and could be fitted into the framework of four basic categories: class, order, genus and species. Every class consisted of at least one order, every order of at least one genus, every genus of at least one species. And vice versa: every species belonged to a definite genus, every genus to a definite order, every order to a definite class. Nature, the treasury of Lord, was seen as a cupboard full of departments, and every department included boxes and these still smaller boxes. A hierarchical order in Nature was not new as such—its idea goes back at least to Aristotle—but the idea of a uniform hierarchy, up to the point of distorting Nature itself, was peculiar to Linnaeus. And—natural or not, theoretically feasible or not—this compartmentalisation of nature was most useful and welcome for all those naturalists who were busily discovering and describing new species; here they had an universal framework where to fit and attach their findings. [46]

Darwin, of course, would come along and kill this view. No longer could all categories be viewed as boxes, as containers with bridgeless gaps.

Because of evolution, categories such as the species category could not be strictly essentialistic. And this is a theme repeated numerous times in philosophy of biology. Darwin killed essentialism in biology. Species taxa are not like chemical elements. A physical atom, of course, can change, by radioactive decay, into an atom of a different element. But one chemical element does not change or evolve into another. Each element, in fact each chemical kind (including compounds such as H_2O), is a static, essentialistic kind. A species taxon, however, since it gradually evolves, or has a new species gradually evolve from it, cannot have a strict essence. Of course clones or monozygotic twins have the same DNA (barring mutations since they individuated), but for the rest of the organisms that constitute a species, variation is the norm, both phenotypic and genotypic (cf. Stamos 2003, 119–122, 143).

And indeed one might argue that in biology such non-Aristotelian categories are now the norm, with concepts such as individual and colony (slime molds being equivalent to dusk and dawn), living and nonliving (ditto for viruses), endo- and ectotherm, adult and infant, male and female, sexual and asexual.

In like manner, that species cannot be essentialistic does not preclude the possibility of an objective category for species. The species category will have species in it and the taxa counted as species will reflect the relevant species concept captured in the particular species definition. The species category has more species taxa in it according to, for example, the species concept of Joel Cracraft than the species concept of Ernst Mayr. The question then becomes the matter of deciding which species concept deserves to be the winner, in the sense of best capturing what a species is. Many think that there should be only one winner, a view known as species monism, the winner being the universal species concept, while others think that biology needs a number of different species concepts, so that the species category should be considered heterogeneous (disjunctive), which is the view known as species pluralism (cf. Ereshefsky 1992b). This is the species problem. If we grant that species are real in nature, then what are they? What is their ontology?

Darwin, as we shall see in subsequent chapters, was not a species pluralist. He did not employ a variety of species concepts. Instead, he repeatedly and consistently employed a set of objective criteria for delimiting species in nature. Moreover, as we shall see in this chapter, Darwin, like many biologists today, thought that there are laws for the species category. He therefore would have to count as a species category realist. This helps to build the case that Darwin was not only a species taxa realist but also a species category realist, with an implicit species concept and definition. Even though, as we have seen in the previous chapter, many have thought, and continue to think, that for Darwin species taxa are arbitrary, we shall see in the following chapters that this view just does not fit the facts. He did not think of species taxa as arbitrary constructs, like constellations, which as pointed out in the previous chapter was the classic example of fictional entities made for convenience and satisfying a convention. Had he done so, then the species category would likewise in his view be a fiction, made simply for convenience. But again, as we shall see in later chapters, Darwin applied repeatedly and consistently a set of objective criteria for delimiting particular species taxa. Moreover, he added that there are objective laws of nature for species *per se*, which by implication means the species category. It is true that Darwin did not *explicitly* make a distinction between species as a taxon and species as a category. Arguably no one did at that time. The distinction does not seem to have been made until after the Synthesis, one of the earliest explicit uses being found in Simpson (1961, 19). But this does not mean that Darwin did not have the distinction in mind. As we have seen at the beginning of chapter 1, he repeatedly employed a distinction between species and "the term species." What we shall see in this chapter is strong evidence that Darwin did not think that the species category is only in the mind.

Sometimes the distinction between taxon and category is phrased using the distinction between group and rank. According to Peter Stevens (1994), "the critical distinction between grouping (forming taxa) and ranking (placement of those taxa in a hierarchy) is rarely made explicit or even implicit in the historical literature" (10). Using George Bentham as an example, Stevens thinks that Bentham was what he (Stevens) calls a "hierarchical nominalist," a group but not rank realist. In Bentham's view, according to Stevens, "groups from variety upwards were more or less discretely bounded, . . . but ranking was arbitrary" (176). Interestingly, Stevens places Darwin in the same category, as a group but not as a hierarchy realist (177). Translated into the language of taxon and category, this is to say that Darwin was a taxon realist but not a category realist. Ghiselin (1969) and Beatty (1985) also claim this for Darwin. I have my doubts that it is true of Bentham (cf. chapter 8), but I am certain that it is not true of Darwin.

For a start, there is something that the group/rank distinction does not capture that the taxon/category distinction does. As pointed out earlier, in some modern views the species category is not just a rank in an existent hierarchy. That view is conformable with the species category merely being a set. Instead, for many, the species category is something more, namely, a subject of laws of nature, which would make it an abstract category, or to vary the terminology again, a *natural kind*.

To get this point across, it is useful to look at the views of Ernst Mayr. More than anyone else, Mayr has stressed the importance of the distinction between species as a taxon and species as a category (e.g., Mayr 1970, 13–14, 1982, 253–254, 1988, 321), and he has likewise stressed the importance of viewing the species category as an abstract class rather than as a set (e.g., Mayr 1970, 13, 1987, 146, 148). But even he has failed to appreciate the view that the species category should properly be thought of as a natural kind. In one of his characteristic discussions on what it is to be a category realist in biology, Mayr (1982) provides two important criteria, which he states as follows: "(1) Are (most of) the groups (taxa) which we rank in the higher categories well delimited? and (2) Is it possible to give an objective (nonarbitrary) definition of such higher categories as genus, family, or order?" (208). Mayr immediately goes on to argue that the species category is real but not taxonomic categories above that level. The important point, however, is that even for species category realism his two criteria, although necessary, are not really sufficient. To be sufficient, I would add the following criterion: (3) Is the category spatiotemporally unrestricted? This criterion seems to me equally necessary, for surely if one holds that there are species taxa but they are only to be found on earth, in spite of holding that life also exists elsewhere in the universe and in an abundance of quantity and complexity, then one could not properly be regarded as a species category realist. To be such a realist, one would agree with Edward Wiley (1981)—though one need not agree with his species concept definition—that "we might expect to find evolutionary species [as he defines the term] anywhere in the universe where organic evolution has occurred" (74). This is in conformity with Dawkins' Universal Darwinism. Anything less than this is not a *category* realism but merely only a *group* (set) realism. The difference, logically, is profound. A category realist would hold that the category (and hence also the definition of the category, providing that it is correct) remains the same even though members of the category come and go in terms of existence. A group realist, on the other hand, cannot make this claim. As members of a group come and go in terms of existence, the definition of the group necessarily changes, since a group is defined by its particular members.

This difference is only further highlighted when we take into account laws of nature (at least on most positive accounts of laws of nature). Categories can have their own laws of nature, but groups *qua* groups cannot. To think otherwise is to deprive laws of their universal character. Indeed it may be said that any positive claim about one or more laws of nature for a particular category is an indubitable hallmark of a category realist, species or otherwise.

As we shall see in the next chapter, Darwin definitely thought that species taxa, at any one horizontal level, are generally well delimited, so he meets Mayr's first criterion. We shall also see, very shortly in this chapter, that Darwin thought that there are genuine laws of nature for species *per se*, so that by my third criterion he thought that the species category is a genuine natural kind. And yet, as we have seen in the previous chapter, he repeatedly stated that "the term species" cannot be defined except arbitrarily, so that, *prima facie* at least, he does not meet Mayr's second criterion.

Given all of this, it would appear that Darwin was seriously muddled on the species question. I do not, however, think that this was the case. Instead, in chapters 3 through 6 I shall attempt to prove that, in spite of his negative claims about defining the term "species," Darwin repeatedly and consistently throughout his mature writings employed a set of criteria for delimiting species. These criteria, moreover, are objective (nonarbitrary) given—as Darwin, of course, was convinced to be the case—evolution primarily by natural selection. What we shall see is that Darwin did in fact have a universal species concept, with an implicit definition. Moreover, in chapter 5 I shall attempt to show that by repeatedly employing this species concept Darwin considered his practice to be in conformity not only with the latest taxonomic advances in other areas of science such as crystallography, but also with the philosophy of science of his day, which put a high premium on laws of nature. Why his species concept and definition remained by him implicit, and that he even repeatedly denied that there could be a definition of the species category, will require an answer in the form of a strategy theory, given in chapter 8, a strategy theory that not only competes with other strategy theories, such as that of Beatty (1985), but that employs to the highest degree the principle of charity as well as conformity to evidence.

For the present, it remains to show that Darwin did indeed think that there are laws of nature for species *per se*, and also to determine as best we can his concept of a law of nature. This, without getting into the details of his species concept (which will be undertaken in subsequent chapters), will lay the groundwork for his species category realism.

For the first task, it will prove to be profitable to begin not with Darwin but with Michael Ghiselin, who has philosophized extensively on the relation between species and laws of nature. Ghiselin is, of course, as pointed out in the Introduction, the father of the modern species-as-individuals view, the view that species taxa are not abstract classes but concrete individuals, individuated both in space and time. Although part of the rationale of this ontological shift was to secure the reality of species taxa against the view of species nominalism (Ghiselin 1966)—since the reality of abstract classes has long been an issue of debate and the claim of their unreality a mainstay of nominalism—it would be a serious mistake to think that for Ghiselin only concrete individuals are real. For Ghiselin the species category is an abstract class and it is just as real. His justification for this conclusion is that there are laws of nature for species, not for individual species taxa but for the species category. (Ghiselin shares the view with many others that there are no laws of nature for individuals but only for classes of individuals.) In other words, the species category in Ghiselin's view is a natural kind. Indeed he devoted a whole paper to this view and articulated what some of these laws (albeit statistical) might be (Ghiselin 1989).[1] And as if to quell any doubt on his species category realism, he reiterated in his book on the metaphysics of species (Ghiselin 1997) that "just in case anybody wonders, it does not seem reasonable to me to say that classes exist only in the mind. If the laws of nature refer to classes, that creates some very sticky problems for anybody committed to such a metaphysical position" (127, cf. 219–230).

What makes this all the more remarkable is that Ghiselin thought both that Darwin was a precursor of the species-as-individuals view and that he was a species category nominalist. With regard to the former, Ghiselin (1969) says "Darwin had proposed a radical solution to the traditional question of the 'reality' of taxonomic groups. What is 'real' is the genealogical nexus, and the groups, or taxa, are chunks, so to speak, of this nexus. As a consequence, it became possible for the nominalist (and Darwin was something of a nominalist) to look upon taxa not as universals but as particulars, or individuals. . . . Thus to Darwin, a taxon is real because it is a clade ('cleft') or genealogical unit" (85). This conclusion, claims Ghiselin, "is demonstrable from his actual procedure in systematic work" (92). It is not important here to repeat or summarize Ghiselin's analysis of Darwin's systematic work, for reasons that should become apparent when I offer my own analysis. What is important here is to notice that Ghiselin interprets Darwin as a precursor of his own view, for which Ghiselin is justly famous (Ghiselin 1974, 1987), the view that species taxa are concrete individuals rather than classes, having both a beginning and an ending in space and time and a horizontal cohesion accomplished by sex. What is also important to notice is that much of Ghiselin's argument for his interpretation of Darwin as a species category nominalist is based on his ascription to Darwin of an Aristotelian concept of definition, such that the reality or naturalness of a category requires a discrete essence. Since in Darwin's view evolution by natural selection precludes such discrete essences, because of its gradualism by minute steps, Darwin was therefore, in Ghiselin's view, a species (and higher) category nominalist. As Ghiselin (1969) puts it, Darwin "insisted on Aristotelian definition as a criterion of reality or naturalness. To Darwin, as to many other taxonomists, an inability to give rigorous definitions for the names of taxonomic groups led to a belief that somehow such assemblages were artificial" (82).

In chapters 3, 4, and 6, I shall argue in detail against Ghiselin's view that Darwin was a precursor of the species-as-individuals view (cf. also Stamos 1998, 1999, 2003, 51–74), also in chapter 6 I shall reconstruct his implicit definition of his mature species concept, while in chapter 8 I shall reply to Ghiselin's claim that Darwin insisted on an Aristotelian concept of category. For the present it is important to point out a serious preliminary problem with Ghiselin's argument, and it has to do with the species category and laws. What is so problematic is not the claim that Darwin was a species taxa realist (although the claim that Darwin thought of them as individuals will be rejected), but that Darwin was also a realist about laws of nature, and not for particular species but for species taxa *per se*. The problem is that if this is true, and we shall see shortly that it is, then the only consistent conclusion based on Ghiselin's own principles is that Darwin was also a species category realist.

Before we look at those examples, however, it is important to examine two laws in particular, accepted by the majority of Darwin's fellow naturalists, which Darwin did in fact reject. His rejection of these laws could easily lead one to conclude that for Darwin the species category is not real. The first law is what might be called the law of limited variation. In the *Origin* Darwin says "It has often been asserted . . . that the amount of variation under nature is a strictly limited quantity" (468). Interestingly Lyell (1832) had listed this in the form of two of his five "laws" for species: "first, that the organization of individuals is capable of being modified to a limited extent by the force of external causes; . . . thirdly, that there are fixed limits beyond which the descendants from common parents can never deviate from a certain type" (23; cf. Darwin 1909, 20 and n. 67). Darwin of course rejected this law in the *Origin*, stating that "the assertion is quite incapable of proof" (468). Of course for Darwin all the evidence he brought to bear in the *Origin* for the fact of evolution seemed sufficient to prove that the law of limited variation must be rejected as a real law.

The second law is closely related to the previous. It is the so-called law of reversion. According to this law, as Darwin in the *Origin* briefly puts it, "domestic varieties, when run wild, gradually but certainly revert in character to their aboriginal stocks" (14). Darwin refers to this law as "well-known" (25), and Wallace (1859) tells us that it had "great weight with naturalists, and has led to a very general and somewhat prejudiced belief in the stability of species" (10). Wallace describes the law as follows: "*varieties produced in a state of domesticity are more or less unstable, and often have a tendency, if left to themselves, to return to the normal form of the parent species; and this instability is considered to be a distinctive peculiarity of all varieties, even those occurring among wild animals in a state of nature, and to constitute a provision for preserving unchanged the originally created distinct species.*" He then adds, with regard to the difficulty that the existence of permanent or true varieties posed for these naturalists, that "the difficulty is overcome by assuming that such varieties have strict limits, and can never again vary further from the original type, although they may return to it, which, from the analogy of the domesticated animals, is considered to be highly probable, if not certainly proved" (10–11). This law had an impressive pedigree. Linnaeus, for example, exhibited a belief in the law of reversion, along with the law of limited variation, in his *Critica Botanica*, first published in 1737, in which he wrote "every day new and different florists' species arise from the true species so-called by Botanists, and when they have arisen they finally revert to the original forms. Accordingly to the former have been assigned by Nature fixed limits, beyond which they cannot go: while the latter display without end the infinite sport of nature" (Ramsbottom 1938, 200 n). Similarly Antoine-Laurent de Jussieu, in the Introduction to his *Genera Plantarum*, first published in 1789, wrote that a species can be "occasionally subverted for a while by chance or human industry; that is to say, some individuals may vary one from another on account of location or climate or disease or cultivation; . . . But these *varieties*, obeying the law of nature, . . . return to the primordial species, their character restored, if other factors do not interfere" (Stevens 1994, 340–341). This law, it is to be noted, was thought to apply only to the species category. Logically there could be no law of reversion for genera and higher taxa, in the minds of those naturalists who held it, not only because they were species fixists, but also because they were higher taxa nominalists (cf. Stamos 2005, 83–84). Darwin in the *Origin*, of course, readily admitted that "In both varieties and species reversions to long-lost characters occur" (473), such as stripes on the shoulders and legs of horses. But his admission only went as far as characters. Of the widespread claim that varieties, if unperturbed, reverted to their primordial state, Darwin in the *Origin* wrote "I have in vain endeavored to discover on what decisive facts the above statement has so often and so boldly been made" (14), and "there is not a shadow of evidence in favour of this view" (15). And again, the law of reversion for Darwin (as for Wallace and their followers) would have to be false if varieties are to be incipient species.

Having rejected two of the most commonly accepted laws for the species category, Darwin in the *Origin* provides a number of his own laws for species, although admittedly some of them are not stated explicitly as laws. What makes them interesting and relevant, nevertheless, is not only that they are implicitly if not explicitly treated as laws by Darwin, but that they also figure as laws in modern evolutionary biology. For a start, we might begin with an apparent anticipation of what is known as the Red Queen hypothesis, named and enunciated by Leigh Van Valen in the early 1970s as a "new law" (Van Valen 1973), according to which a species must evolve just to maintain its place in nature and to avoid extinction, since its physical environment is continually changing as well as the species, due to competition, with which it interacts. In the *Origin* Darwin states "Hence we can see why all the species in the same region do at last, if we look to wide enough intervals of time, become modified; for those which do not change will become extinct" (315). The statistical nature of this law is evident in the fact that there exist "anomalous forms [which] may almost be called living fossils; they have endured to the present day, from having inhabited a confined area, and from having thus been exposed to less severe competition" (107).

Darwin similarly appears to have anticipated what is known as Dollo's Law, named after the version of the law enunciated by Louis Dollo in the late 1880s, according to which the statistical probability is extremely low that evolution will ever precisely reverse or repeat itself (cf. Gould 1970). In the *Origin* Darwin states "When a species has once disappeared from the face of the earth, we have reason to believe that the same identical form never reappears" (313), the reason being "for the link of generation has been broken" (344), which he further elaborates as follows: "We can understand why a species when once lost should never reappear, even if the very same conditions of life, organic and inorganic, should recur. For though the offspring of one species might be adapted (and no doubt this has occurred in innumerable instances) to fill the exact place of another species in the economy of nature, and thus supplant it; yet the two forms—the old and the new—would not be identically the same; for both would almost certainly inherit different characters from their distant progenitors" (315). Granted, Darwin in all of this is mainly talking about groups, for he says, following the second quotation, "When a group has once wholly disappeared, it does not reappear, for the link of generation has been broken (344), and following the third quotation, "Groups of species, that is, genera and families, follow the same general rules in their appearance and disappearance as do single species, . . . A group does not reappear after it has once disappeared" (316). Nevertheless, that Darwin applies the law to species is highly significant given, as we have seen and shall see in subsequent chapters, that he thought of species as real.

To the above one might add that for Darwin the extinction of species is also a law of nature. Although this does not come out explicitly in his discussion on extinction in the *Origin* (109–111, 127, 172, 317–322, 433, 475), it is nevertheless highly implicit and is made explicit in one of his letters to Andrew Murray (April 28, 1860), in which he states "the extinction of every form of life in the course of time is a law of nature" (Burkhardt *et al.* 1993, 179).

Darwin also thought it a law of nature that each species has a single center of origin in both space and time, as opposed to multiple origins, even for situations (contra, e.g., Louis Agassiz, James Dana, and Alphonse de Candolle) where you have "exceptional cases of the same species, now living at distant and separate points," as Darwin put it in the *Origin*. Given the origin of species by means of natural selection and the various means of geographic distribution, the latter discussed by Darwin in Chapter XI of the *Origin*, Darwin thought that single origins constitute a "universal law" (354).

Of course the most important law for Darwin, of which all the above may be said to be in large part derivative, is natural selection, the main mechanism for Darwin of adaptation and species evolution. For a start, right at the end of the Introduction to the *Origin* Darwin tells us that he is "convinced that Natural Selection has been the main but not exclusive means of modification" (6), to which we need to keep in mind the title of his book: *On the Origin of Species by Means of Natural Selection*. However, rather than proffer natural selection in the *Origin* as a law of nature, Darwin calls it a "principle" (61, 80, 95, 127, 475), a "theory" (237, 245, 320, 322, 472, 469), and a "doctrine" (95), as well as a "process" (109, 169, 467) and a "power" (61, 109, 205, 242, 454, 469). Of course the distinction needs to be made between the *principle* or *theory* or *doctrine* of natural selection (which is man-made and revisable) and the *process* or *power* of natural selection (which is in nature), the former being better or worse depending on how well it captures (describes) the latter (assuming the latter exists). The same applies to laws. There needs to be a distinction between the *laws of science* (which are man-made and revisable) and the *laws of nature* (which if they exist are irrespective of the former) (cf. Weinert 1995,

4–5). Nevertheless, it remains a fact that nowhere in the *Origin* does Darwin explicitly call natural selection, whether the principle/theory/doctrine or the process/power, a "law." Instead, he leaves it implicit, as when, for example, he refers to "one general law, leading to the advancement of all organic beings, namely, multiply, vary, let the strongest live and the weakest die" (244), or when he refers to "the laws which have governed the production of so-called specific forms" (472), or when he says "these elaborately constructed forms, so different from each other in so complex a manner, have all been produced by laws acting around us. These laws, taken in the largest sense, being Growth with Reproduction; Inheritance which is almost implied by reproduction; Variability from the indirect and direct action of the external conditions of life, and from use and disuse; a Ratio of Increase so high as to lead to a Struggle for Life, and as a consequence to Natural Selection, entailing Divergence of Character and the Extinction of less-improved forms" (489–490). Achingly close, but still no cigar. Equally tantalizing is that in his correspondence at the time he wrote the *Origin* Darwin repeatedly stressed that natural selection is a bona fide *vera causa* (true cause). For example, in a letter to Charles Bunbury (February 9, 1860) he wrote "With respect to Nat. Selection not being a 'vera causa'; . . . Natural selection seems to me in so far in itself not to be quite hypothetical, in as much as if there be variability & a struggle for life, I cannot see how it can fail to come into play to some extent" (Burkhardt *et al.* 1993, 76). In Darwin's day, the claim that natural selection is a *vera causa* would be taken as equivalent to the claim that natural selection is a law, more specifically a causal law such as gravity. Darwin himself indicates this in one of his letters to Lyell (February 23, 1860), in which he wrote "With respect to Bronn's objection that it cannot be shown how life arises, & likewise to certain extent Asa Gray's remark that natural selection is not a vera causa.— I was much interested by finding accidentally in Brewster's life of Newton, that Liebnitz objected to the law of gravity, because Newton could not show what gravity itself is" (Burkhardt *et al.* 1993, 102; cf. also Darwin's letters to Gray, February 24, 1860, and Lyell, June 17, 1860, Burkhardt *et al.* 1993, 106–107, 258). Elsewhere in Darwin's writings, however, well before and after this period, the label is explicit, as when in the *Essay of 1844* (Darwin 1909) he says, when comparing artificial to natural selection, "Very differently does the natural law of selection act" (95), or when in his autobiography (1876a) he says "The old argument of design in nature, as given by Paley, which formerly seemed to me so conclusive, fails, now that the Law of Natural Selection has been discovered" (87).

In chapter 5 we shall examine more clearly in what sense Darwin thought natural selection a law of nature, as well as what he thought of the nature of laws in general. Moreover we shall look closely at how he connected natural selection as a *vera causa* with the *versa causa* ideal in the philosophy of science of his time. In arguing that natural selection is the *vera causa* of species, Darwin was bringing evolutionary biology within the circle of genuine science, which at the time was circumscribed philosophically by an emphasis on laws of nature. What others did for crystallography, for example, Darwin did for biology. At any rate, for the present it is sufficient to conclude, before getting into the specifics of Darwin's species concept, that he was indeed a species category realist.

Chapter 3

The Horizontal/Vertical Distinction
and the Language Analogy

If one surveys the modern literature on the species problem, one will find a number of distinctions that together define the debate. Two, of course, are realism versus nominalism and species as a category versus species as a taxon, both of which have been dealt with in the previous two chapters. Two others are species as classes versus species as individuals and species as process entities versus species as pattern entities, both of which we shall deal with in this and in later chapters. But there is a further distinction, which is just as common as the others although it is not always explicit. This is the temporal distinction between species as horizontal entities versus species as vertical entities.

This distinction borrows from the now entrenched evolutionary metaphor of the Tree of Life, instituted, of course, by Darwin (1859, 129–130). Basically, horizontal species exist as one cuts through the Tree horizontally. The idea here is not necessarily one of species existing at an instant or microsecond. Instead, the idea usually is that of a time frame smaller than what is minimally needed for the completion of gradual speciation. This is the view of species, to name but two examples, contained in the biological species concept of Mayr (1942, 273, 1970, 12, 1991, 186), arguably still the dominant species concept in zoology, and in the morphological species concept, which arguably is the dominant species concept in botany (cf. Cronquist 1978; Luckow 1995; McDade 1995). What needs to be added and emphasized is that the conception of species as primarily horizontal entities need not preclude their existence over vast stretches of time. Instead, it is only to affirm that their reality is primarily horizontal.

Vertical species, on the other hand, are species conceived either as minimal (branchless) branches on the Tree, as one finds in cladistics (e.g., Hennig 1966, 58–66; Ridley 1989, 3), or as differentiated segments of a minimal branch, as one finds in the theory of punctuated equilibria (e.g., Gould 1982, 109–110; Eldredge 1985, 115, 122). Both of these views are subsumed under the ontological paradigm of species that has reigned in recent decades, namely, the species-as-individuals view first argued by Ghiselin (1966, 1974, 1997) and then Hull (1978, 1988). In this view, a species is not an abstract class but instead a spatiotemporally restricted physical individual, having a beginning in time (speciation), an ending in time (extinction), and a definite location in space (Ghiselin and Hull add integration by sex, but some of their followers do not; cf. the references in Stamos 2003, ch. 4), with membership in a species being a part/whole relationship. Moreover since species, in this view, are genuine individuals very much like individual organisms, they cannot possibly exist again once they have gone extinct. As with you or me, once we are dead, we are not just dead but most sincerely dead, and later copies of us (no matter how perfect) would not be you or me.

Although some have been reluctant to admit the horizontal/vertical distinction (e.g., Laporte 1994, 158 n. 49; de Queiroz 1999, 53–54), most who contribute to the literature on the species problem in some way recognize it. Interestingly, a number of biologists, although their terminologies differ, have used the distinction to not only distinguish their species concepts from others but even to dichotomously classify competing species concepts. Simpson (1961), for example, characterizes species according to Mayr's biological species concept as "contemporaneous" and "horizontal" (164), while species according to his own evolutionary species concept are "lineages" (153) with "a long time dimension" (154). Similar to Simpson, who characterizes the debate as "species at any one time" versus "species delimited in a certain span of time" (166), Mayr (1988) characterizes the debate as "whether an evolutionary (vertical) concept is superior to the standard (horizontal) biological species" (314). In a similar vein, Salthe (1985) divides modern species concepts into "diachronic" and "synchronic" (225–226), Endler (1989) uses "contemporaneous versus clade" (627), while Ridley (1993) uses "non-temporal" and "temporal" (385) and, synonymously, "horizontal" and "vertical" (399).

The distinction also relates to the issue of species realism versus nominalism. Nothing makes this clearer than contrasting two quotations from Mayr and Eldredge. According to Mayr (1988), "Modern biologists are almost unanimously agreed that there are real discontinuities in organic nature, which delimit natural entities that are designated as species" (331). According to Eldredge (1985), however, "Indeed, at any moment, seemingly at least half the world's evolutionary biologists are perfectly prepared to deny that species—*any species*—even exist" (98). These quotations look like they are mutually inconsistent, but they are not really. Mayr, the ornithologist, is a neontologist, meaning that he is used to dealing with living (or recently living) organisms. Not surprisingly his biological species concept is a horizontal species concept. When he says that most biologists recognize the reality of species, then, he is referring to species as horizontally conceived, and his claim makes sense given that most modern biologists are neontologists of one sort or another and that when viewed horizontally it is fairly easy to argue that species are real (cf. Stamos 2003, 80–97). Eldredge, on the other hand, is a paleontologist, used to dealing not only with fossilized organisms in horizontal strata but also with their trends through strata. Not surprisingly his species concept, which is based on his theory of punctuated equilibria, is a vertical species concept. When he says that at least half of all biologists think that species are not real, he is thinking of species as vertical entities. And indeed, although most biologists are neontologists, most do think that if viewed vertically, species are unreal (cf. Stamos 2003, 76–79). Determining whether species are (or should be conceived as) primarily horizontal or vertical entities, therefore, is a matter of some importance.

At any rate, if Darwin was truly a species realist, and a thoughtful one at that, we should expect to find evidence that he conceived of species taxa either as primarily horizontal or as primarily vertical entities. And we do. But which was it? Ghiselin (1969), as we have seen in the previous chapter, claimed that Darwin was a precursor of the species-as-individuals view. But only a precursor. In addition he claimed "Darwin seems not to have fully appreciated the importance of discontinuity, and in this sense he did not embrace what we would call a biological species concept. However, he did recognize that there are species, and he did conceive of them as units or stages in the evolutionary process. Perhaps the evolutionary species concept of Simpson is the closest modern parallel to Darwin's" (101). Malcolm Kottler (1978), on the other hand, while following Ghiselin's view that Darwin was a species taxa realist though not a species category realist, thought that the closest modern parallel was Mayr's biological species concept. He claims that in the *Origin* and later writings Darwin "did appreciate the importance of biological criteria, including reproductive isolation, as opposed to purely morphological criteria, in the characterization of species and did believe that species were in some sense real" (292). Moreover he makes much of Darwin on sibling species in willow wrens (279–280 and n. 9), though at the end of his paper he waters it down by adding "One might say that for Darwin in his later works this relationship between interbreeding and transition was not so clear—closer to 'less' than 'more' acknowledged" (297).

In chapter 6 I shall argue against Kottler's claim about Darwin on reproductive isolation, although in the present chapter we shall see that he was right if his view was indeed that Darwin (like Mayr) thought of species as primarily horizontal entities. As a consequence we shall also see that Ghiselin's interpretation of Darwin does not withstand scrutiny. If Darwin was a precursor of the species-as-individuals view, if his view on species was closer to Simpson's view than anyone else, then we should be able to find strong evidence that Darwin thought of species as primarily vertical entities. But all the evidence, as we shall now see in this chapter, points to the opposite conclusion, to the conclusion that Darwin thought of species as primarily horizontal entities. We shall also explore some of the interesting implications of this.

Before we get into Darwin's mature writings, however, it is important to notice that at the beginning of his career as an evolutionist he did seem to toy with the idea of species as individuals. This was when he held what today is called his "monadic theory of evolution" (cf. Gruber 1981, 103, 129–149). Three of the essential ideas here were a belief in spontaneous generation (which of course was fairly widespread), a belief in a fixed duration of time for the existence of each species, and a branching conception of life. Accordingly in his early notebooks (Barrett *et al.* 1987) we find entries such as "Tempted to believe animals created for a definite time:—not extinguished by change of circumstances" (Red Notebook, 129). This was written in about March 1837 (Barrett *et al.* 1987, 18), while contemplating what he believed to be an extinct species of llama. Again in the same notebook he says "There is no more wonder in extinction of species than of individuals" (133). Similarly in Notebook B, begun about July 1837, he wrote "There is nothing stranger in death of species, than individuals. If we suppose monad definite existence [= definite duration], as we may suppose is the case, their creation being dependent on definite laws, then those which have changed most must in each state of existence have shortest life; Hence shortness of life of Mammalia" (22–23).

The idea that each species has a predetermined lifespan would not be new to Darwin. The Italian geologist Giovanni Brocchi, for example, suggested in 1814 that every species is constituted in such a way as to have a predetermined lifespan, so that a species must eventually go extinct even if its environment remained favorable. Darwin would have learned of Brocchi's theory from his reading of the second volume of Lyell's *Principles of Geology* (1832), which he read while onboard the *Beagle*. As Lyell put it, "The death, he [Brocchi] suggested, of a species might depend, like that of individuals, on certain peculiarities conferred upon them at their birth" (128), a theory that Lyell immediately went on to reject although he held Brocchi high in his estimation of him as a geologist.

At any rate, for reasons that are not entirely clear (cf. Gruber 1981, 140–145), Darwin quickly gave up the monadic theory of evolution (ejecting spontaneous generation and the aging thesis but retaining the branching conception of life), which had a lifetime of maybe four months (having been born just before the summer of 1837 and dying in September of that year). During that brief period he began to believe that the extinction of a species is not internally predetermined but contingent on external circumstances. As he put it in Notebook B, "death of species is a consequence (contrary to what would appear from America) of non adaptation to circumstances" (38–39). There was possibly in this change of view some occasional backsliding (cf. Notebook B, 135; Gruber 1981, 145), but if so they were death spasms. Darwin never again entertained the idea of species aging.[1]

And indeed it is possible, in spite of the analogies made in his monadic theory of evolution, that Darwin never even thought of species at this time as individuals, as having a beginning and ending in space and time. Interestingly, however, Darwin's son Francis has claimed otherwise. Quoting passages from Notebook B—specifically, "Propagation explains why modern animals same type as extinct which is law almost proved" (14), "They die; without they change; like Golden Pippens, it is a *generation* of *species* like generation of *individuals*" (63), and "If *species* generate other *species*, their race is not utterly cut off" (72)—Francis Darwin (1909) claims that these quotations "show, I think, that he recognized the two things not merely as similar but as identical" (xii).

Interestingly, the idea of species as individuals was a bit in the air at the beginning of the twentieth century. We can also see the idea in the first book by Francis Darwin's friend and younger contemporary Julian Huxley (grandson of T.H.). For a start, Huxley (1912) writes of "the species-individuality of which we are the parts" (24), he makes a distinction between two kinds of individuality, one for organisms and one for species (25), he claims that "If evolution has taken place, then species are no more constant or permanent than individuals" (27), and he adds "As individual emerges from individual along the line of species, so does species emerge from species along the line of life" (27–28).

In spite of Francis Darwin, however, I suggest that if Darwin did indeed think of species as individuals, it was only during the brief stint of his monadic theory of evolution. Even then, there is strong evidence that he thought of species primarily as horizontal entities, not as vertical entities as required by the species-as-individuals view.

In chapter 1 we have seen that in the transmutation notebooks Darwin employed something very close to the modern biological species concept. What is interesting is that, in accordance with this view, he thought of species as primarily horizontal. As we have seen in chapter 1, in one of his definitions of species he says, in Notebook C, "As species is real thing with regard to contemporaries—fertility must settle it" (152). This is a later annotation to what he wrote on that page, specifically "A species is only fixed thing with reference to other living being.—one species May have passed through a thousand changes, keeping distinct from other & if a first & last individual were put together, they would not according to all analogy breed together." Already here we can see, in league with his pre-modern version of the biological species concept (the modern version of which is adamantly a horizontal species concept), an emphasis on the primarily horizontal reality of species (cf. Hodge 1987, 241). What adds to this is an increasing reliance on a different analogy for species, a reliance away from that of the individual organism and toward that of the individual language.

The shift to the language analogy is a matter that I shall take up later in this chapter, including its important philosophical consequences. For the present, I want to turn to the time of the *Origin* and show that Darwin clearly conceived of species as primarily horizontal entities.

We have seen in chapter 1 that in reply to Agassiz's quip, that if species are not real they cannot vary, Darwin wrote that species have a "temporary" existence. But what did he mean by "temporary"? From an evolutionary point of view, whether species are conceived of as primarily horizontal or vertical, they are in each case "temporary." I believe the evidence is overwhelming, however, that by "temporary" Darwin had "horizontal" in mind.

To begin, in the *Origin* Darwin says "To sum up, I believe that species come to be tolerably well-defined objects, and do not at any one period present an inextricable chaos of varying and intermediate links" (177;

cf. 203). This is clearly a reference to the horizontal dimension, and equally clearly its tone is that of species realism.

On the other hand, only two pages later, Darwin says "Lastly, looking not at any one time, but to all time, if my theory be true, numberless intermediate varieties, linking most closely all the species of the same group together, must assuredly have existed" (179). Here we see that even though Darwin is dealing with a long time dimension, there is no talk of species vertically conceived, only of species linked by intermediate varieties. In fact Darwin used a number of phrases in addition to "intermediate varieties," namely, "transitional forms," "transitional varieties," and "intermediate links," all meaning the same thing. For example, in a passage about the fossil record, Darwin claims that all throughout biological history there must have been "an infinite number of those fine transitional forms, which on my theory assuredly have connected all the past and present species of the same group into one long and branching chain of life" (301). In a section devoted to "the absence or rarity of transitional varieties," Darwin says "extinction and Natural Selection will, as we have seen, go hand in hand. Hence, if we look at each species as descended from some unknown form, both the parent and all the transitional varieties will generally have been exterminated by the very process of formation and perfection of the new form" (172). In the passage following this, Darwin uses the phrase "innumerable transitional forms." Finally, this time in his concluding chapter, on the imperfection of the geological record, Darwin says "The number of specimens in all our museums is absolutely as nothing compared with the countless generations of countless species which certainly have existed. We should not be able to recognise a species as the parent of any one or more species if we were to examine them ever so closely, unless we likewise possessed many of the intermediate links between their past or parent and present states; and these many links we could hardly ever expect to discover, owing to the imperfection of the geological record" (464).

In all the above, species are clearly horizontal species. We do not find the idea of species commonly extending through a long stretch of time, let alone changing—and least of all changing radically—and yet remaining the same species. In fact, nowhere in Darwin's writings do we find this. Darwin's primary conception of species was horizontal.

Some comments are necessary, however, before we continue with the evidence. First, by "well-defined . . . at any one period" Darwin did not mean that he thought species are well-defined in the sense of essential characters as captured in an Aristotelian definition. This meaning is eliminated by the fact that Darwin conceived of species evolution as minutely gradual, which makes it impossible for a species to have an essence (cf. Stamos 2003, 122). Instead, by "well-defined" Darwin was simply following common usage, in which the phrase means sufficiently clear or well marked out and thereby allowing for delineation and description. We can see this usage especially well near the end of the *Origin*, wherein Darwin says "Systematists will only have to decide (not that this will be easy) whether any form be sufficiently constant and distinct from other forms, to be capable of definition; and if definable, whether the differences be sufficiently important to deserve a specific name" (484; cf. also 171, 174, 432, 1871 I, 226–227). Moreover the phrase "well-defined" for species seems to be used synonymously by Darwin with "good and distinct species" (259) and with "good and true species" (47). This usage is common even today. For example, Mayr and Short (1970) state that "The designation 'good species' refers to those that are clearly delimited from other species and whose recognition is not controversial among specialists" (1). Moreover there was nothing idiosyncratic in Darwin's use of the term "definition" for "description" even in his own time. Hugh Strickland (1837), for example, in his preliminary suggestions for rules for zoological nomenclature, wrote "A name may be expunged which has never been clearly defined. . . . Many collectors of shells and fossils are in the habit of labelling those species which they do not find described, with names of their own invention; but, unless they publish descriptions of these new species, they cannot expect these names to stand" (174). Whether by Darwin, by some his contemporaries, or even by modern biologists, then, there is nothing in the above phrases to imply essentialism.

Second, we should not suppose that by "at any one time" Darwin meant an instant or microsecond for horizontal species. In fact in the passage above where Darwin affirms the horizontal reality for species, he uses the words "at any one period." Darwin clearly uses both phrases with the same reference. Moreover when we look at Darwin's comparison of species with languages, which he conceives of both as primarily horizontal, the only plausible interpretation is that Darwin did not conceive of the horizontal dimension as an instant, but rather as a period of time which, at its maximum, is too short, in the case of species, for the gradual evolution of one species into another. Today this is thought to be roughly 3,000 years (cf. Stamos 2002, 177). In the case of languages, of course, where language is defined traditionally as a system of communication, the maximum time period for the horizontal dimension is going to be much shorter.

Third and finally, we have to keep in mind that, among his few correspondents who were evolutionists, one in particular seemed to conceive of species as primarily horizontal. As we have seen in chapter 1 (note 4), the botanist H.C. Watson, in a letter to Darwin (November 8, 1855), wrote "I must confess a pretty strong bias towards the view, that species are *not* immutably distinct;—altho' in our time-narrowed observation of the individuals they seem to be so" (Burkhardt and Smith 1989, 499). Similarly, a little later in a letter to Darwin (December 20, 1857) he wrote "I cannot find the proof of species being definite & immutable, whatever they may seem to be at any one time & spot" (Burkhardt and Smith 1990, 511).

But why would Darwin think that "at any one period" species in general are "tolerably well-defined objects"? Immediately following the passage in the *Origin* in which this quotation is found, he gives us three reasons. The first is that variation is a very slow process and natural selection requires uninhabited niches, the second is that intermediate varieties will often be exterminated by natural selection—this connects with his view that "extinction and natural selection . . . go hand in hand" (172)—and the third is that intermediate varieties, because of their smaller numbers, will often be exterminated by accidents (presumably such as fire etc.) (177–178).

To these three reasons given by Darwin in the *Origin*, we also have to add that he thought that interspecific hybridization is comparatively rare (ch. 8). A further consideration is that Darwin thought that most species have a fairly well-defined range. As Darwin put it, "In looking at species as they are now distributed over a wide area, we generally find them tolerably numerous over a large territory, then become somewhat abruptly rarer and rarer on the confines, and finally disappearing. Hence the neutral territory between two representative species is generally narrow in comparison with the territory proper to each" (174). This view, incidentally, accords with what modern biologists write (e.g., Mayr 1970, 301–302; Futuyma 1986, ch. 13). Of further importance is Darwin's pluralism with regard to speciation. Darwin allowed for both allopatric (geographic isolation) and sympatric (principle of divergence) modes of speciation (more on this below). Both modes entail branching speciation and his metaphor of the "great Tree of Life" (130). As he says in the *Origin*, "In a tree we can specify this or that branch, though at the actual fork the two unite and blend together" (432). It is only the forks, then, at any one horizontal dimension, which present messy situations, situations without good and distinct species. But since the forks, at any one horizontal dimension, are not the norm, most species at that dimension are going to be good and distinct species.

It is important here to further understand Darwin's theory of speciation. According to the influential theory of Mayr (e.g., 1982, 410–417), although Darwin was an early adherent of speciation by geographic (allopatric) isolation, by the time he wrote the *Origin* he no longer had a clear theory on speciation, so that his book was misnamed. A more likely scenario, it seems to me, is that Darwin evolved into a limited speciation pluralist (limited by his confinement to gradualism),[2] accepting by the time of the *Origin* both nonbranching speciation or anagenesis (cf. 119–120) and branching speciation or cladogenesis, the latter including both sympatric speciation (in accordance with his principle of divergence; cf. 105–106, 112, 115–116) and allopatric speciation (cf. 399–404). Of course, that he considered sympatric speciation the most important (cf. 105) does not take away from this pluralism, which, by the way, seems to be a growing view in modern speciation studies (cf. Otte and Endler 1989; Howard and Berlocher 1998).

Another objection might be based on Darwin's work on barnacles. Darwin had spent eight years anatomizing and monographing barnacles from 1846 to 1854, both living and fossil. Did not their variability prove to him that species (or rather what were called "species") are too variable, even at any one horizontal level, to be real, so that (under his breath at least while writing his monographs) he gave up species as real entities? The answer is no. Granted, Darwin often complained of their variability. For example, in a letter to Hooker (June 13, 1850) he wrote "I have been struck (& probably unfairly from the class) with the variability of every part in some slight degree of every species: . . . & consequently that the diagnosis of species from minute differences is always dangerous. . . . Systematic work w^d be easy were it not for this confounded variation, which, however, is pleasant to me as a speculatist though odious to me as a systematist" (Burkhardt and Smith 1988, 344). Moreover there is that often-quoted passage in another of Darwin's letters to Hooker (September 25, 1853) in which he wrote, in reference to his barnacle work, "After describing a set of forms, as distinct species, tearing up my M.S., & making them one species; tearing that up & making them separate, & then making them one again (which has happened to me) I have gnashed my teeth, cursed species, & asked what sin I had committed to be so punished" (Burkhardt and Smith 1989, 156).[3]

In spite of this "confounded variation," however, Darwin nevertheless found that barnacles on the whole, when studied anatomically, presented good species. In a letter to Lyell (September 2, 1849), for example, Darwin remarked "I sometimes after being a whole week employed & having described, perhaps *only* 2 species agree mentally with L^d. Stanhope that it is all fiddle-faddle" (Burkhardt and Smith 1988, 252, italics mine). Instead, the "chaos" that Darwin so often complained about with regard to barnacle systematics referred only to the "book" species of his contemporaries, which were based only on external characters and which often involved a nomenclature with many synonyms. For example, in a letter to Hugh Strickland (February 4, 1849), on the topic of the rules of nomenclature, Darwin wrote "In not one large genus of Cirripedia has *any one* species been correctly defined: . . . Literally not one species is properly defined: not one naturalist has ever taken the trouble to open the shell of any species to describe it scientifically, & yet all the genera have ½ a dozen synonyms. . . . The subject is heart-breaking" (Burkhardt and Smith 1988, 206–207). Similarly in a letter to Johannes Müller (February 10, 1849) he wrote "I find that the anatomy of the Cirripedia has been most imperfectly done; nearly all the most striking features in their organization having been overlooked.—Their classification is likewise a perfect chaos, as must be the case until the whole body of every species be examined, as I am now doing" (Burkhardt and Smith 1988, 213).

And in fact when we look specifically at Darwin's work on living barnacle species we find that he claims that most, when properly anatomized, are good species. For example, in his preface to his second volume, on the Lepadidae (Darwin 1851b), he wrote that "no doubt they are subject to considerable variation, and as long as the internal surfaces of the valves and all the organs of the animal's body, are passed over as unimportant, there will occasionally be some difficulty in the identification of the several forms, and still more in settling the limits of the variability of the species" (xi). Even when he dealt with the most problematic cases of variation, Darwin still maintained that, when properly anatomized, most species will turn out to be good species. For example, in his second volume on living species, on the Balanidae (Darwin 1854b), in his introductory discussion to the genus *Balanus*, which for Darwin was by far the largest genus and which he considered to have "an especial amount of variation" (156), he remarked that "Notwithstanding the difficulties now enumerated, I hope that, owing to having examined a vast number of specimens of the most varying species, I have not fallen into many errors" (190). Surely Darwin would not have felt this way if he thought that most of the species he carefully anatomized were not after all good species. Moreover one has only to read his descriptions of his 45 species of *Balanus* (194–302), notwithstanding his additional 9 species of the subgenus *Acasta* (302–321), to see his confidence in regarding most of the species as sufficiently good and distinct.

Of the latter subgenus, granted, Darwin says the first four species caused him "much doubt and trouble" (308), while with species 5 and 6, though very distinct from other species, he also expresses "some hesitation" (314) in making them specifically distinct. With species 7 he doesn't say, while species 8 is "perfectly distinct" (318) and species 9 is "very distinct" (319).

With *Balanus* the situation is much better, in spite of its variability. Granted, species 1 and 19 are described as "the most difficult and variable in the genus" (196–197) and as causing Darwin "utter despair" (243). Species 18 is described as causing Darwin "much trouble" (235). With species 34, 35, and 36, taken together, Darwin feels "somewhat doubtful" (284). Similarly with species 43 and 44 Darwin has "some doubt whether they ought to be specifically separated" (298). With species 4 and 8 he expresses a little difficulty. Species 6, however, "can easily be distinguished" (213). Species 10, 11, 12, and 13 are described as "quite distinct species" (218). Species 17 is described as a "well-marked species" (232). Species 20 is described as having "amply diagnostic characters" (247). With species 21 and 22 Darwin says "I entertain no doubt whatever about the distinctness of the two species" (250). Species 23 is described as a "very distinct species" (253), as is species 30 (274) and 38 (288). Species 24 is described as differing "distinctly" from species 27 (254). Species 25 is described as being "very distinct from every other" (259). Species 33 is described as being a "distinct and well-defined species" (281). Species 27 and 29, although "confounded" in most collections (261), are found by Darwin, when "disarticulated," to be "at once distinguished," which is only further confirmed by Darwin's observation that they inhabit different depths of water (271). Similarly species 32, when "disarticulated," "cannot be confounded with any other" (278). Similarly as well, when species 45 is "disarticulated" the result allows "of no mistake of the two species" (302), the other species being species 27. Species 37 is described as being "so peculiar" (285). Species 41 is described as being a "strongly characterized species" (294). Species 42 presents "well defined distinctions" (295). Finally, with species 2, 3, 5, 7, 9, 15, 16, 26, 28, 31, 39, and 40, it is implied that they are good species.

To all of the above, one needs to keep in mind that *Acasta* and *Balanus* were the most difficult groups.

So we can see that in the *Origin* Darwin was not disingenuous when he claimed that horizontally most species are good species. Moreover it is abundantly evident that the horizontal dimension was for him the primary dimension for species reality. But this is not to deny that Darwin allowed for a species, horizontally conceived, to exist vertically over a vast stretch of time. For a start, he coined the term "living fossils." These "anomalous forms," he says, which he later calls "species and groups of species" (486), "have endured to the present day," and the reason he gives for their existence, along with the name, implies that they have remained virtually the same over vast stretches of time to the present, the reason being "from having inhabited a confined area, and from having thus been exposed to less severe competition" (107). He also tells us that "species very rarely endure for more than one geological period" (153). Moreover when we look at Darwin's work on fossil barnacles (Darwin 1851a, 1854a), in spite of the problem that the soft parts were generally not preserved in the fossil record, we find that in a number of cases Darwin classified fossil species as conspecific with living species. In the case of *Balanus* (1854a, 13–33), which as we saw was divided by Darwin into 45 living species, Darwin recognized 11 fossil species, each of which he identified with a living species, and in each case because the diagnosis seemed to be the same.

All of this will make more sense when we examine what Darwin thought was the closest analog of species, namely, natural languages. But what needs to be reiterated at this point is that in allowing a species to have a vertical extension this did not mean that Darwin thought that species are primarily vertical. We have already seen much evidence that Darwin thought of species as primarily horizontal. Later in this chapter I shall argue that Darwin was on firm logical ground. What I want to do at this point is provide a further piece of evidence that for Darwin the reality of species is primarily horizontal. This further piece of evidence comes from Darwin's discussion on his one and only diagram in the *Origin* (the diagram is reproduced on the next page). Interestingly, according to Kevin de Queroz (1999)

> An early version, or at least a precursor, of the general lineage concept can be found in Darwin's (1859) *Origin of Species*. In the only illustration in that book, Darwin represented species as dashed and dotted lines, or collections of such lines, forming the branches of what would now be called a phylogenetic tree. In the accompanying text, he used the term *species* more or less interchangeably with the term *lines of descent*. [76]

In this passage de Queroz reads Darwin as having a vertical species concept, conceiving of species as branches in the phylogenetic tree of life. But this reading does not at all survive close scrutiny of what Darwin actually said in relation to his diagram. Perhaps de Queroz was misled by the concluding section of the chapter containing the diagram, in which Darwin states "The affinities of all the beings of the same class have sometimes been represented by a great tree. I believe this simile largely speaks the truth. The green and budding twigs may represent existing species; and those produced each former year may represent the long succession of extinct species" (129). But even here species are compared to buds, not branches. Instead a branch represents a "long succession of extinct species."

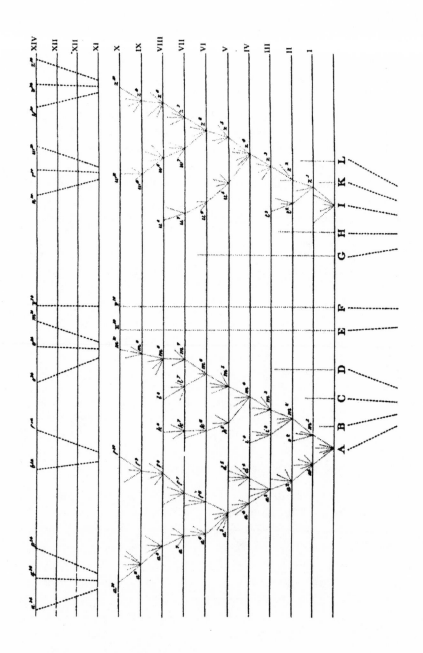

This interpretation only becomes stronger once we turn to Darwin's discussion on his diagram. It is interesting, and surely not insignificant, that F^{14}—descended from species F, says Darwin, 14,000 generations later, or better yet 114,000 generations later, "either unaltered or altered only in a slight degree"—is twice called by Darwin a "new species" (117, 124). Certainly no one motivated by a primarily vertical species concept would call F^{14} a new species in such a diagram. Simpson (1961) certainly did not do this sort of thing in his own diagrams (164, 168). Neither would cladists, since they conceive of species as branches in the phylogenetic tree of life, and species conceived as such are allowed to undergo unlimited change and remain numerically the same just so long as they don't produce branches (cf. Hennig 1966, 58; Ridley 1989, 10). (I should add that the line from species F to F^{14} in Darwin's diagram is a straight line with no branching; the same is true of the line from E to E^{10}.) Nor is there any evidence that Darwin thought of species in the way adherents of the modern theory of punctuated equilibria do, as vertical entities characterized mainly by stasis with rapid evolution (which equals for them speciation) delimiting the temporal ends (cf. Eldredge 1985, 115, 122). Even though, as we have already seen, Darwin did countenance stasis, he did not use it to conceive of species, which would follow alone from his belief in the relative infrequency of living fossils.

Although Darwin's diagram indicates yet once again that for him the horizontal dimension has priority over the vertical dimension, it should not be taken to necessarily preclude the vertical dimension, as indicated by what he says about living fossils. Moreover, it should be noticed that the F in F^{14} is capitalized (unlike, e.g., a^{14} or f^{14} or m^{14}), which alone suggests that F^{14} alone on that horizontal line may be thought of as a vertical species numerically identical with its corresponding species on the horizontal base line, namely, species F (and so also for E^{10} with E), unlike the lines for the lowercase letters which seem to indicate only lineages and not species.

To understand Darwin's thinking further on this matter it will help to turn to his main analogy for species, namely, the language analogy. We have seen near the beginning of this chapter that Darwin held a monadic theory of evolution very briefly at the beginning of his career as an evolutionist. In this theory a species is a vertical entity highly analogous to an individual organism. Darwin quickly gave up that theory and the organism analogy, however, replacing it with a theory of contingent evolution by natural selection and the language analogy for species. This analogy is highly significant and philosophically of the utmost importance.

Beginning with Darwin's early notebooks (Barrett *et al.* 1987), in Notebook B he wrote "As man has had not time to form good species, so cannot the domesticated animals with him!—Modern origin shown by only one species, far more than by non-embedment of remains—?agrees with non-blending of languages?" (244). In Notebook N he wrote "many learned men seem to consider there is good evidence in the structure of language, that it was progressively formed" (65). In Notebook OUN he wrote "At least it appears all speculations of the origin of language.—must presume it originates slowly—if these speculations are utterly valueless— then argument fails—if they have, then language was progressive.—We cannot doubt that language is an altering element, we see words invented— we see their origin in names of People.—Sound of words—argument of original formation.—declension &c often show traces of origin" (5). He also wrote "H. Tooke has shown one chief object of language is prompt- ness of consequence hence languages become corrupt, & whole classes of words are abbreviations he thus derives from nouns & verbs—so that much of EVERY language shows traces of anterior state??" (13). In these passages it is not always entirely clear whether Darwin is commenting on the evo- lution of the human language ability or on the evolution of languages. In- deed as Stephen Alter (1999) puts it, "These passages start out as speculations regarding the origin of speech, yet they quickly shade into re- flections on the purely analogic similitude between languages and species" (16). Although the two ideas are commingled, they nevertheless do show that Darwin was starting to think of languages as analogous to species. What is especially interesting is what he says in the last quotation above. John Horne Tooke was a British linguistic philosopher who argued that French, Italian, Anglo-Saxon, Dutch, German, Danish, Swedish, and English "are little more than different dialects of one and the same lan- guage" (Barrett *et al.* 1987, 603 n. 13–1). In capitalizing the word "EVERY" in the quotation above, we can see that Darwin, against Tooke, held a commonsense view of language, which may legitimately shed light on what Darwin meant by the word "language" in the previous passages (and possibly also in later passages) where he compares species to languages.

In the *Sketch of 1842* there are allusions to the language analogy as used by Lyell for geology, which I shall return to below. Of far more interest is the *Essay of 1844* (Darwin 1909), in which Darwin compares the impor- tance of rudimentary organs for determining classification of species in the natural system with rudimentary structures in languages. As Darwin put it, "In the same manner as during changes of pronunciation certain letters in a word may become useless in pronouncing it, but yet may aid us in searching for its derivation, so we can see that rudimentary organs, no longer useful to the individual, may be of high importance in ascertaining its descent, that is, its true classification in the natural system" (235).

This is an idea that Darwin would repeat in the *Origin* (455). There, however, we find more comparisons between species and languages. We are told, for example, that "a breed, like a dialect of language, can hardly be said to have a definite origin" (40). Equally important, in arguing that the only natural classification of species is genealogical, Darwin proceeds to "illustrate" his view "by taking the case of languages," of which he concludes "the proper or even only possible arrangement would still be genealogical; and this would be strictly natural, as it would connect together all languages, extinct and modern, by the closest affinities, and would give the filiation and origin of each tongue" (422–423).

But it is in the *Descent of Man* (1871 I) that we find the most striking passages. Therein Darwin writes:

> The formation of different languages and of distinct species, and the proofs that both have been developed through a gradual process, are curiously the same. . . . The frequent presence of rudiments, both in languages and in species, is still more remarkable. . . . Languages, like organic beings, can be classed in groups under groups; and they can be classed either naturally according to descent, or artificially by other characters. Dominant languages and dialects spread widely and lead to the extinction of other tongues. . . . The same language never has two birth places. Distinct languages may be crossed or blended together. We see variability in every tongue, and new words are continually cropping up; but as there is a limit to the powers of memory, single words, like whole languages, gradually become extinct. . . . The survival or preservation of certain favoured words in the struggle for existence is natural selection. [59–61]

Much of what is interesting about Darwin's language analogy is his sources. In the final chapter of the first volume of his *Principles of Geology*, Lyell (1830, 461–462) compares change in geologic epochs to change in a written language over a period of 1,000 years. In both cases, he says, if one knows only one-tenth of the objects of study as they exist today, one cannot possibly hope to have a good understanding of the objects of study in the distant past. In the third chapter of the third volume of his *Principles*, Lyell (1833, 33–34) supposes that if an archaeologist were to find two buried cities at the foot of Mount Vesuvius, one on top of the other, and found the inscriptions at the lower city to be Greek and the inscriptions at the higher city to be Roman, he would infer incorrectly if he thought the language change from Greek to Roman was abrupt. And if he later found a third city in between the two, with inscriptions in Italian, he would then see his error, and infer properly that the language change from the earlier population to the later had been "very gradual."

Darwin had read all three volumes of Lyell's *Principles* assiduously during his *Beagle* voyage (Burkhardt and Smith 1985, 562), and he would soon after consider himself the heir to Lyell's uniformitarianism, as applied to the evolution of species. He was also familiar with a letter that John Herschel had written to Lyell on February 20, 1836 (extracted in Charles Babbage's *The Ninth Bridgewater Treatise*, published in 1837), in which Herschel wrote "Words are to the Anthropologist what rolled pebbles are to the Geologist— Battered relics of past ages often containing within them indelible records capable of intelligent interpretation," and in which he called the origin of species the "mystery of mysteries" (cf. Darwin 1859, 1; Burkhardt and Smith 1986, 8–9 and n. 5; Desmond and Moore 1992, 214–215).

Many years later, in his *The Antiquity of Man*, Lyell (1863) devoted a whole chapter (ch. 23) to the comparison between language evolution and the theory of species evolution and made a number of remarkable comparisons. Ontologically speaking, the most important of his comparisons are that "the learned are not agreed as to what constitutes a language as distinct from a dialect" (458), that practically speaking "two languages should be regarded as distinct whenever the speakers of them are unable to converse together" (458), which he compares to the sterility barrier, that there is a "real question . . . whether there are any limits to . . . variability" (462), that an important area of inquiry is over language change, in particular "what are the laws which govern not only the invention, but also the 'selection' of some of these words or idioms, giving them currency in preference to others" (463); that "dialects . . . may be regarded as . . . 'incipient languages'" (464), that "we find in them some internal evidence of successive additions by the invention of new words or the modification of old ones" (465), that "No one of them can have had two birth places" (465), that "They may die out very gradually in consequence of transmutation, or abruptly by the extermination of the last surviving representatives" (467), and that "a language which has once died out can never be revived, since the assemblage of conditions can never be restored" (467). Some of these analogies are straight from Darwin, as we have seen, but they surely prompted Darwin to strengthen the analogy even further in his *Descent*, again as we have seen. Interesting is what Darwin wrote in 1863 in a letter to Gray (February 23), that "Lyell was pleased, when I told him lately that you thought that language might be used as excellent illustration of derivation of species; you will see that he has *admirable* chapter on this" (Burkhardt *et al.* 1999, 166), and what he wrote to Lyell (March 6) that "No praise can be too strong, in my opinion, on that inimitable chapter on language in comparison with species" (207).

Of greater interest is what was going on in the field of linguistics during Darwin's lifetime (the term "linguistics" was introduced to England by Whewell 1837). While Darwin was developing his evolutionary views, and most biologists thought of species as fixed (allowing at most for varieties), virtually all linguists thought of languages as evolutionary. Interestingly, what became prominent near the beginning of the nineteenth century, particularly in Germany, was the organism metaphor. This vertical conception of languages occurred hand in hand with the concern for determining particular paths of language evolution and with reconstructing ancestral languages, most of all, Proto-Indo-European. Notable in this regard were Friedrich von Schlegel and Jacob Grimm, both of whom in the first decade of the ninteenth century claimed that the discipline closest to comparative grammar was comparative anatomy. Of much greater importance was Franz Bopp, highly influential as the author of the first comparative grammar of the Indo-European languages. In 1827 Bopp wrote "Languages must be taken as organic natural bodies which form themselves according to definite laws, develop carrying in themselves an internal life principle, and gradually die" (Davies 1987, 84). Bopp, however, arguably did not take the organism metaphor literally (cf. Wells 1987, 56). Nevertheless, he seems to have thought, as Davies (1987) points out, that "a language can change its grammatical 'type' while remaining in some sense the 'same' language" (91). At any rate, the greatest exponent of the organism metaphor was August Schleicher, regarded by many as the most influential figure in ninteenth century linguistics (cf. Taub 1993, 175–176 and n. 13). As early as 1848 Schleicher began to argue that individual languages *really are* organisms, imbued with life in a literal sense so that the scientific study of them is a biological science in its own right, up there with zoology, botany, and paleontology. In 1863, in finding further support for his views upon reading Heinrich Bronn's German translation of Darwin's *Origin*, he wrote "Languages are natural organisms that arose independently of human volition, grew in accordance with definite laws, and developed and in turn grew old and died. They are also characterized by the set of phenomena we are accustomed to understand by the term 'life.' Accordingly, glottics, the science of language, is a natural science; its method is largely the same as that of the other natural sciences. This is why the study of Darwin's book . . . was bound to seem quite close to my area" (Percival 1987, 8). Equally important, it was

Schleicher who made famous the use of branching tree diagrams to illustrate genealogical relationships in language evolution. Moreover, he argued that any study of language that was not evolutionary was not scientific (cf. Taub 1993, 188).

As Alter (1999) points out, "The 'new philology' of Bopp and Grimm began seeping into England in the early 1830s, in part through the efforts of Darwin's cousin and brother-in-law Hensleigh Wedgwood" (11). One of my main criticisms of Alter's otherwise excellent book (Stamos 2000) is that he doesn't go far enough in telling us what Darwin got from Wedgwood. Granted, Darwin had many discussions with Wedgwood during his stay in London following his *Beagle* voyage, Wedgwood's main focus in his own work was on etymology and the natural origin of language (which would of course have been of interest to Darwin), Wedgwood had written an extensive review (Wedgwood 1833) of Grimm's *Deutsche Grammatik*, and they kept up with correspondence and visits after Darwin moved to Down. And yet we are not told anything about the organism metaphor. Gillian Beer (1989) provides much more on Wedgwood's review of Grimm, and tells us that "It is likely that Darwin read his cousin's only published article, given his current interest in the subject and the close ties between them" (158). And yet, in spite of the pages devoted by Beer to Wedgwood's review (158–160), we are not told anything about the organism metaphor. Beer (157) also claims that Darwin had probably read Alexander Hamilton's review (Hamilton 1820) of Bopp's *Conjugations Systems*. And yet again we are told nothing of the organism metaphor. In fact, upon examining Hamilton (1820) and Wedgwood (1833) myself, I found not the slightest mention or even hint of the organism metaphor. Nevertheless, given Darwin's close connection with his cousin Hensleigh, given the scientific milieu of his time, and given Darwin's wide reading, it is highly unlikely that he would not have known about the organism metaphor that had taken over linguistics.

Granting this, Darwin arguably resisted the organism conception of languages, and like Lyell thought of languages in what might rightly be called the commonsense view, as primarily horizontal entities delimited by intercommunication. This follows from his comparison of species with languages, which, with regard to the former, we have seen was primarily horizontal. Granted, although Darwin's main analog for species became languages after he relinquished his monadic theory of evolution, one can still find him once in a while comparing species with organisms. But the comparison is only with regard to the production by secondary laws. For instance, in the *Sketch of 1842* (Darwin 1909) he wrote "It accords with what we know of the law impressed on matter by the Creator, that the creation and extinction of forms, like the birth and death of individuals should be the effect of secondary means" (51). Similarly in the *Origin* he wrote "To my mind it accords better with what we know of the laws impressed on matter by the Creator, that the production and extinction of the past and present inhabitants of the world should have been due to secondary causes, like those determining the birth and death of the individual" (488). But the occasional uses of such comparisons does not mean that Darwin thought of species as primarily vertical entities, in the manner of individual organisms. Darwin's comparison was causal, not ontological.

Moreover there was a precedent for this in linguistics. Wilhelm von Humboldt, friend and mentor of Bopp and brother of Alexander von Humboldt (whose *Personal Narrative of Travels* had an enormous influence on Darwin's choice to become a scientist), arguably maintained a primarily horizontal conception of languages while others were adopting the vertical conception, although this did not prevent him from using organism metaphors and analogies for languages (cf. Davies 1987, 101–103, nn. 38 and 45; Harris 1993, 19). Moreover, Darwin had a strong reason for not making the comparison between species and organisms a strong one, and that was principally because in his view a species, unlike an individual organism, does not have a developmental program. Indeed this is why Darwin refused to call his theory of evolution a "developmental hypothesis," which was the name often given to the theories of Lamarck and Chambers.[4] As Darwin put it in a letter to Leonard Jenyns (February 14, 1845), "Thanks for your hints about terms of 'mutation' &c; I had had some suspicions, that it was not quite correct, & yet I do not see my way to arrive at any better terms; it will be years before I publish, so that I shall have plenty of time to think of better words— Development w^d. perhaps do, only it is applied to the changes of an individual during its growth" (Burkhardt and Smith 1987, 143). And indeed Darwin had twice in the *Origin* made clear that he did not subscribe to what would later be called *orthogenesis* (the theory, prevalent in the early 20th century, that each species has a developmental program built into it carrying it in a direction of change independent of selection imposed by the environment). In the *Origin* he says "I believe in no fixed law of development, causing all the inhabitants of a country to change abruptly, or simultaneously, or to an equal degree. The process of modification must be extremely slow. The variability of each species is quite independent of that of all others. Whether such variability be taken advantage of by natural selection, and whether the variations be accumulated to a greater or lesser amount, thus causing a greater or lesser amount of modification in the varying species, depends on many

complex contingencies" (314; cf. 351). Similarly, in a letter to Hugh Falconer (October 1, 1862) on the topic of natural selection, Darwin wrote "I
suspect that you mean something further,—that there is some unknown law
of evolution by which species *necessarily* change; and if this be so, I cannot
agree" (Burkhardt *et al.* 1997, 441). For the same reason Darwin resisted the
word "evolution" in the *Origin*, because of its use for the development of an
individual organism ("ontogeny" today, the Latin *evolutio* and *evolvere* meaning "to unfold," as in unfolding a scroll), preferring the phrase "descent with
modification" instead. (I shall reply to Robert Richards on this matter in
Chapter 10.) And again, Darwin did not take up William Harvey's suggestion (October 8, 1860) that Darwin replace "Natural *Selection*" with "Natural *Evolution*" as expressing "the combination of all the powers of nature in
the production of species" (Burkhardt *et al.* 1993, 416).

Interestingly, linguistics eventually came around, following the revolution brought to a head by Ferdinand de Saussure roughly at the turn of the
century, switching from what he called *diachronic* linguistics (which more
often than not had previously involved the organism metaphor) to what he
called *synchronic* linguistics, in which the ontology of a language is considered as a system of communication and accordingly a primarily horizontal
entity. This revolution has stuck (for details, cf. Stamos 2002, 178–183).
There is today, following the Copernican revolution of Saussure, in spite of
robust health in the field of historical linguistics, no *cladistic* language concept. There is no view of language in professional linguistics as expressed by
Ghiselin (1997), according to whom "English has never undergone the ana-
logue of speciation. The language of *Beowulf,* the *Canterbury Tales, Hamlet,*
and *Huckleberry Finn* has changed a great deal without ceasing to be one
and the same individual language" (141). Rather, the modern idea is that,
keeping with English and as Quine (1986) put it, "later English," in contradistinction to "earlier English," is "another language" (14).

Darwin did not have the vertical conception of language, as indeed nor
did Lyell. Arguably his conception of language was the commonsense one,
that of a primarily horizontal entity, what Fodor (1975) in an entirely different context called "the good old way: viz., as a system of conventions for
the expression of communicative intentions" (106). All of this was in spite
of what was going on in the field of linguistics around him. And as we have
seen, Darwin thought the same of species, not as individuals, not as vertical entities, but as strongly analogous to languages commonly conceived.

And if indeed I am correct in this, it further seems to me that Darwin
was on firm logical ground in placing the reality of species primarily in the
horizontal rather than in the vertical dimension. There are a number of

arguments that drive home this point, each of them good. John Dupré (1981), for example, argues that "the objective reality of the [phylogenetic] branch can be no greater than the objective reality of the grouping of organisms that constitutes the beginning of the branch" (88). Since any of a number of different relations between organisms, or between organisms and their environment, could be used to define the group of organisms which in turn constitutes the branching point, the vertical perspective alone is dependent on the horizontal. As Dupré further puts it, "the phylogenetic criterion must be parasitic on some other, synchronic, principle of taxonomy. It cannot generate privileged properties on its own" (89). Elsewhere in my own writings (Stamos 1996, 138 n. 36, 1998, 462, 2003, 79), I argued that, for species as with languages, the horizontal dimension is logically and therefore ontologically prior to the vertical dimension for the simple reason that a horizontal species can be real even though it lacks a vertical reality (if it evolves gradually one cannot at all say, even roughly, when it began and when it came to an end), while a vertical species cannot possibly be real without having successive horizontal realities. In my paper devoted specifically to the topic (Stamos 2002), I was required to provide a further argument in order to satisfy an anonymous referee and the editor. There I used an *argument by analogy*, arguing that because species are much more analogous to languages than to individual organisms, and languages according to the modern view (and rightly so) are primarily horizontal entities, it follows that the ontology of species is *probably* also primarily horizontal (*probably* because argument by analogy can never establish anything more than a high probability). The nature of the argument can be schematized as follows (l = any natural language, s = any species, o = any organism, H = primarily horizontal, V = primarily vertical):

> l and s both have properties *ABCDEFG*
> l has the property H
> Therefore, s has the property H with a probability of p
>
> s and o both have the properties *TU*
> o has the property V
> Therefore, s has the property V with a probability of q
>
> $p > q$
> Therefore, it is more probable that s has the property H than the property V

It remains to be argued that species and languages are similar in more relevant respects than species and individual organisms. The following list (which is greatly simplified from Stamos 2002, 186–191), though not exhaustive, will suffice: (i) Species and languages are plastic in a way that organisms are not. Change in an organism is constrained by its genotype. Species and languages have no such constraint. They have phylogeny, but not ontogeny. (ii) With species as with languages, horizontally variation is the norm. With a multicellular organism, on the other hand, although there may be much variation in cell types, it is superficial given the unity of the genome shared by all the cells. (iii) With both species and languages, variation does not present the same problems horizontally as it does vertically. Horizontally, both species and languages, although they often have geographic variation, have in most cases fairly clear borders or ranges, beyond which there are relatively few individuals, allowing for species maps and language or dialect maps. Viewed vertically, however, the only borders one can appeal to are branching points, which present their own kind of problems given that the history of life, and even more so the history of languages, is far from being perfectly hierarchical. These features and problems have no analogs to change within an individual organism. (iv) Species and languages exhibit both anagenesis (unidirectional) and cladogenesis (branching) patterns of change, as well as both sympatric and allopatric (including founder effect) modes of formation. The ontogeny of an individual organism, on the other hand, is anagenetic only superficially, since it retains basically the same genotype throughout its lifetime, while its modes of reproduction only superficially resemble sympatric and allopatric models of speciation and language formation (even with binary fission and budding, the favorite analogs of species-as-individuals theorists, there is nothing analogous to mitosis in amoeba reproduction or in the copying of the large and small replicons in bacterial reproduction). (v) Species and languages are subject to similar forces and mechanisms driving their change, namely, random variation, natural selection, and drift (especially in founder populations and bottlenecks). Although these forces and mechanisms are not completely the same in species and languages, they nevertheless have much more in common than the forces or mechanisms that drive change in an individual organism (vi). With individual organisms, it is unequivocal that multiple births do not result in numerically the same individual. Instead it produces siblings. Even if the multiple births are qualitatively identical, the result is not numerical identity. A copy of me, no matter how similar, is not *me* (e.g., if it goes out and commits murder, *I* am not guilty). For species and languages on the other hand, it is not at all unequivocal that multiple births result in multiple offspring. In plants, for example, repeated allopolyploidy between two parental species is routinely treated by botanists as resulting in a third species, not a third, fourth, and fifth, etc., new species. In other words, the multiple origins, the resulting allopoly-ploid plants or populations, are routinely treated as conspecific, as being all of one species (e.g., Ashton and Abbott 1992). If the same were to happen in linguistics, no modern linguist would treat the different groups as speaking numerically different languages, if their languages were qualitatively identical, because they in principle could communicate with each other. In species biology, however, those committed to a cladistic or species-as-individuals view are committed to identifying the multiple origins as numerically distinct species, even though they are qualitatively identical (or near identical). This odd consequence has been highly criticized, and rightly it seems to me. At any rate, it is avoided once species are taken to be primarily horizontal. (vii) In a closely related matter, while organisms typically undergo an enormous amount of change from birth to death, it is unequivocal that such protean change results in numerically the same individual throughout. With species and language change, however, it is not unequivocal. If the language of a population on an isolated island were to undergo radical change over time, so that in principle one of the language users at a later date could not possibly understand one of the language users at an earlier date (and *vice versa*), no modern linguist would say they speak one and the same language. The result is a *lineage* of language systems, not a language system. In species biology, on the other hand, some biologists (again cladists, proponents of species-as-individuals) want to maintain that if a species undergoes "infinite evolution," then it is still numerically the same species throughout (e.g., Ridley 1989, 10). But many biologists do not think this way (e.g., Mayr 1970, 248). (viii) With organisms, it is unequivocal that each organism ages. In linguistics, that view for languages died over 100 years ago (cf. Fox 1995, 33 n. 3). In species biology, the idea is similarly dead. Not only does a species lack a genetic program and consequently the genetic mechanisms of aging, but as Raup (1991) also pointed out, a species that stays above its *minimum viable population* number is "virtually immune to extinction" (124, cf. 6). (ix) Finally, with individual organisms, it is unequivocal that the death of an individual is necessarily (logically) forever. Again, should a copy of me be made after I die, it is not *me*. With species, however, while cladists and proponents of species-as-individuals claim that the extinction of a species is necessarily forever (they balk at the thought of *Jurassic Park*), many other biologists entertain the view that the extinction of a species is not necessarily forever—Richard Dawkins (1986, 73) is one example, and Darwin, as we shall see in the next chapter, was another. In modern linguistics, although the revival of a "dead" language is considered highly improbable, it is not precluded on logical grounds (cf. Trask 1996, 329–330). And this is because languages are viewed horizontally, as systems. Systems are not fixed and timeless entities, but they are not spatiotemporally bounded either. Rather they are dynamic entities that are logically free to re-evolve and re-exist whether at the same time and in different places or at different times and places. They are genuine historical entities, but of a very different sort than individuals.

This view of species has interesting implications for speciation. Of course it is meaningless to speak of "speciation" unless one has first defined what one means by "species." We have not yet done that for Darwin, but in determining that species for him are primarily horizontal, we can indeed draw some implications for speciation just from the horizontal nature of species alone.

First, speciation is coupled to evolution. A species that undergoes "infinite evolution" is not a species but something else, what might be called a "species lineage" (cf. Stamos 2002, 190).

Second, it is sometimes thought that since the horizontal dimension is static, speciation is impossible, so that speciation needs a vertical dimension. But the horizontal dimension is not static. Instead it is dynamic, gradually changing with every day, and as such provides (in the sense of time) everything that is needed both for species reality and for speciation (cf. Stamos 2003, 324).

Third, the concept of species as primarily horizontal entities is not a concept of chronospecies. This is a concept from paleontology, which refers to chopping up a lineage in the fossil record into successive species. Chronospecies are typically held to be arbitrary, ultimately nothing more than subjectively delimited segments of a continuum, mere matters of taxonomic convenience. And of course from a vertical perspective, chronospecies are indeed arbitrary. From a horizontal perspective, however, vertical species appear just as arbitrary! Thus, given a phyletic lineage in the fossil strata that spans, say, 10 million years, the question "How many horizontal species, on your view, are there?" is an illegitimate question, since it is asked from the vertical perspective. From that perspective, one might just as well answer that there are infinitely many species as that there are none. The answer makes no sense because the question was asked from the wrong perspective, wrong because the vertical dimension is neither logically nor ontologically the primary dimension for species reality. The concept of chronospecies is a product of that wrong perspective, and that is what makes it an illegitimate, nonsensical concept. It is not the horizontal dimension that makes it so. From that dimension, at any one horizontal level it makes perfect sense to ask how many species there are, and from that perspective meaningful questions get meaningful answers (cf. Stamos 2003, 316–317).

In arguing for species as primarily horizontal entities, I am definitely going against a modern trend, namely, the trend in the past few decades toward a cladistic or phylogenetic conception of species in biology and the corresponding species-as-individuals view in philosophy of biology. However, as I also pointed out (Stamos 2002), one can see the beginnings of a reverse trend in phylogenetic systematics, a trend away from a vertical and back to a horizontal species concept.

Two examples should suffice. The first is the phylogenetic species concept of Joel Cracraft. Originally conceiving of species as individuals (Eldredge and Cracraft 1980, Cracraft 1987), he more recently (without notice) changed his view, stating (Cracraft 1997) that species in his view are not clades (branches) (326) but "terminal taxa" (332), a term in phylogenetic systematics that refers to the tips of clades. Another example is the genealogical species concept of David Baum and Kerry Shaw (1995), which is based on coalescing gene trees and according to which "exclusive groups of organisms are only meaningfully delimited among groups of organisms living at any one point of time" (300). (For more on these species concepts, cf. Stamos 2002, 193–194, 2003, 270–276, 278 n. 68.)

I believe that Darwin, could he have known of this today, would have embraced this reverse trend, and for the reasons I have given I would say that in doing so he would be right.

Chapter 4

Common Descent
and Natural Classification

It might be thought that Darwin's emphasis on genealogy for classification, on common descent, completely undermines the analysis in the previous chapter. For example, in the *Origin* Darwin repeatedly emphasized that the "Natural System" of classification, so debated and desirously sought after by his fellow naturalists, is in fact a genealogical one. As he put it, "all true classification is genealogical; that community of descent is the hidden bond which naturalists have been unconsciously seeking, and not some unknown plan of creation, or the enunciation of general propositions, and the mere putting together and separating objects more or less alike" (420). Does not Darwin's emphasis on genealogy entail that he thought of taxa, both species and higher, as vertical entities, as branches in the Tree of Life?

What only aids this supposition is that many biologists today, irregardless of Darwin, think that if species are to be evolutionary entities, real entities in nature rather than timeless abstract entities, then they must be thought of as vertical entities, indeed as individuals. Ghiselin (1974) and Hull (1978), for example, have routinely argued that everything real is either a class or an individual, classes are timeless entities, species evolve, therefore species must be individuals. (I have inveighed against the class/individual distinction as a false dichotomy in Stamos 1998, 456–459, 2003, 214–215.) Following Ghiselin and Hull, a number of authors have claimed that if species are historical entities then they must be individuals. For example, Eldredge (1985) claims that "punctuated equilibria puts the icing on the cake in the argument that species are real historical entities, comparable in a formal manner to individual organisms" (122). Moreover, one can find the related claim that the importance of history reconstruction requires a vertical species concept. Frost and Wright (1988), for example, claim that "only individuals (= entities) exist independently of definition and have histories. Thus, only individuals should be included in biological taxonomies" (201).

Returning to Darwin, in the *Origin* he seems to be saying the same thing when he says "If we extend the use of this element of descent,—the only certainly known cause of similarity in organic beings,—we shall understand what is meant by the natural system: it is genealogical in its attempted arrangement, with the grades of acquired difference marked by the terms varieties, species, genera, families, orders, and classes" (456). To this we may add what he says near the very end, *viz.* "when we regard every production of nature as one which has had a history . . . how far more interesting, I speak from experience, will the study of natural history become!" (485–486). Moreover, this was a view that Darwin long held. For example, in a letter to George Waterhouse (July 26, 1843) he wrote "According to my opinion . . . classification consists in grouping beings according to their actual *relationship*, i.e. their consanguinity, or descent from common stocks . . . . it is clear that neither number of species—nor grade of organization ought to come in, as an element" (Burkhardt and Smith 1986, 375–376). In *Descent* (1871 I) Darwin states, with regard to the natural system, that "This system, it is now generally admitted, must be, as far as possible, genealogical in arrangement. . . . The amount of difference between the several groups—that is, the amount of modification which each group has undergone—will be expressed by such terms as genera, families, orders, and classes" (188).

By focusing on common descent for natural classification, it might seem that Darwin went so far as modern taxonomists of the now dominant cladistic variety, who insist that taxa (both species and higher) must be *monophyletic*. This latter term generally has two meanings. The more general sense is that of single origin, meaning that a taxon cannot have multiple origins (as in repeated polyploidy). In phylogenetic systematics, *polyphyletic* groups are not considered natural groups. The other, more strict sense of monophyly (which is not always applied to species but always to higher taxa), is that of a taxon consisting of the common ancestral organisms and all descendant organisms. If some of the descendant organisms are not included in the group, then the group is called *paraphyletic*, and most (but not all) of those who subscribe to phylogenetic systematics eschew paraphyletic taxa since it brings in the criterion of similarity.

When we take a closer look at Darwin, we see that in stressing common descent he did not at all come close to modern cladistics. In the case of Darwin's systematic work on barnacles, Ghiselin and Jaffe (1973) show that Darwin went some way toward a phylogenetic classification but by no means as far as cladism, since "he definitely did have paraphyletic taxa in his system" (138), and also because he included amount of difference and other practical considerations (134, 139). I suggest that when looking at what Darwin did with other groups of organisms we get an even better picture of his view on genealogy and classification, including especially as it pertains to the species category.

First, it is important to notice that, unlike species, Darwin did not think that higher taxa are real. For example, early on, again in a letter to Waterhouse (July 31, 1843), Darwin wrote "I believe . . . that if every organism, which ever had lived or does live, were collected together (which is impossible as only a few *can* have been preserved in a fossil state) a perfect series would be presented, linking all, say the Mammals, into one great, quite indivisible group—and I believe all the orders, families & genera amongst the Mammals are merely artificial terms highly useful to show the relationship of those members of the series, *which have not become extinct*" (Burkhardt and Smith 1986, 378). To this he added that "classification . . . is governed by the breaks or chasms in the series" (378). We can see the same view implied in the *Origin*, where Darwin says "I believe that the *arrangement* of the groups within each class, in due subordination and relation to the other groups, must be strictly genealogical in order to be natural; but that the *amount* of difference in the several branches or groups, though allied in the same degree in blood to their common progenitor, may differ greatly, being due to the different degrees of modification which they have undergone; and this is expressed by the forms being ranked under different genera, families, sections, or orders" (420).

But how strongly did Darwin insist on common descent when it came to *species*? It would seem rather strongly but not totally.

A good example to begin with is the case of domestic dogs. Virtually all naturalists from Buffon onward (as is the case with biologists today) categorized all the breeds of domestic dogs as members of a single species, *Canis familiaris*, because of their interfertility. However, it is striking that nowhere in his writings, either in his publications, his notebooks, or his correspondence, did Darwin follow this practice. In other words, there is nowhere that Darwin calls domestic dogs a single species. The main reason, quite apparently, is because he believed that the group as a whole was domesticated not from a single wild species but from a number of wild species, domestication eventually eliminating the sterility barriers between the domesticated animals from the wild species. This issue often came up in Darwin's correspondence, with Charles Lyell being almost always his correspondent on this issue (cf. Burkhardt and Smith 1991, 357, 362–364, 384, 386, 392, Burkhardt *et al.* 1993, 170, 258, 261, 262–263, 320, 335, 366, 378, 383–384, 393, 397, 399), who by this time had come to think, contrary to his earlier view (Lyell 1832, 27–28), that all domestic dogs are descended from wolves.

In the *Origin*, Darwin's refusal to classify domestic dogs as a single species is even more evident, in particular alongside the related case of domestic cattle, namely, European cattle and humped Indian cattle. Explicitly following (in part) the doctrine of Pyotr Pallas, Darwin wrote "I believe, for instance, that our dogs have descended from several wild stocks; yet, with perhaps the exception of certain indigenous domestic dogs of South America, all are quite fertile together; and analogy makes me greatly doubt, whether the several aboriginal species would at first have freely bred together and have produced quite fertile hybrids. So again there is reason to believe that our European and the humped Indian cattle are quite fertile together; but from facts communicated to me by Mr. Blyth, I think they must be considered as distinct species" (254). To add to this, early in the *Origin* we find Darwin state that "I should think, from facts communicated to me by Mr. Blyth, on the habits, voice, and constitution, &c., of the humped Indian cattle, that these had descended from a different aboriginal stock from our European cattle; and several competent judges believe that these latter have had more than one wild parent" (18).

Thus the same conclusion, on Darwin's view, follows (though implicit) for domestic dogs, which Darwin discusses, as analogous, immediately above the cattle example, namely, that in spite of their perfect interfertility they must be classified as more than one species because of their multiple origin and differentiated characters (cf. also *Variation* 1868 I, 21, 26, 31–32, 34).

I will explore the significance of differentiated characters for Darwin's species concept in the next chapter. For the present, it is interesting to focus more closely on Darwin's source, namely Edward Blyth. In the *Origin* Darwin tells us that Blyth is one "whose opinion, from his large and varied stores of knowledge, I should value more than that of almost any one" (18). Given such high esteem for Blyth, we should trace the source from Blyth for Darwin's reference to humped Indian cattle. In a letter to Darwin (April 21, 1855), written in India where Blyth was residing from 1841 to 1862 as the curator of the museum of the Asiatic Society of Bengal, Blyth wrote "It is easy to dogmatize & say that *Bos indicus* is but a humped variety of *B. taurus*, but the stump is one of many differential characters, though seen very early in foetal life!" (Burkhardt and Smith 1989, 312). The differential characters of *Bos indicus*, in spite of hybridizability with *Bos taurus*, include not only the muscular hump, but a great difference in voice, in habits such as never seeking shelter from the sun, and in never standing knee and belly deep in water. In a later letter of the same year (September 30 or October 7), Blyth adds a further number of morphological differences, such as differences in height, ears, and horns. He then states "What more can be required to characterize it as a peculiar species? Why the advocates of the opposite opinion can adduce only the prolificacy of the hybrids" (451). Aside from the significance of Blyth's species concept (which quite apparently is not based on fertility or sterility but on constant and distinct characters), what is significant for our immediate purpose is that in these and other letters Blyth nowhere suggests a multiple origin for the two cattle forms. Instead it is Darwin who infers that and, at least in part, uses the lack of common descent to keep them separate as species.[1]

Indeed the topic is only highlighted by Darwin's view on convergence. In a letter to Darwin (January 3?, 1860), H.C. Watson suggested, as a correlate to Darwin's principle of divergence, that Darwin allow "the hypothesis or inference that individuals *converge* into orders, genera, species, as well as *diverge* into species, genera, orders, through nepotal descent" (Burkhardt and Smith 1989, 11–12). In reply to Watson (January 5–11), Darwin stated "With respect to 'convergence' I daresay, it has occurred, but I should think on a very limited scale . . . and only in case of closely related forms. . . . the same cause acting on two closely related forms (i.e. those which closely resembled each other from inheritance from a common parent) might confound them together so closely that they would be (*falsely* in my opinion) classed in same group" (18).

The possibility of convergent evolution, which we see here Darwin allows, raises an interesting issue about similarity. We have seen earlier in this chapter that Darwin claims that "propinquity of descent" is "the only known cause of the similarity of organic beings" (413; cf. 456). And yet convergent evolution is also a real cause of similarity, and Darwin recognizes this. Indeed he distinguishes the two kinds of similarity by calling the former "real affinities" and the latter "analogical or adaptive resemblances" (427). The latter are adaptive because animals "belonging to two most distinct lines of descent, may readily become adapted to similar conditions, and thus assume a close external resemblance." An obvious example is that of whales and fish. The problem for Darwin, however, is that convergent similarity is "almost valueless to the systematist," the reason being that "such resemblances will not reveal—will rather tend to conceal their blood-relationship to their proper lines of descent" (427). Interestingly Lyell, in a letter to Darwin (October 28, 1859), raised much the same problem for Darwin's theory of the polyphyletic origin of domestic dogs, in that "if the three wild species mentioned above ['wolf, fox & jackal'] could produce any or all these races of dogs (or incipient species) by coalescing & interbreeding, I find it impossible to hope to trace any clue to the past transformations of species or their probable birthplaces" (Burkhardt and Smith 1991, 363). At any rate, in *Descent* (1871 I) Darwin comments on the extreme unlikelihood, calling it "utterly incredible," of convergence at the species level "as to lead to a near approach to identity throughout their whole organisation" (231).

Thus far it seems that Darwin's species concept has a strong element of common descent in it, strong enough perhaps to make it a primarily vertical conception of species. However, in addition to what we have seen in the previous chapter, there is further evidence that leads away from this conclusion. Most important is that Darwin did not think that extinction is necessarily forever. Granted, in the *Origin* he can be found saying in his chapter on geology "When a group has once wholly disappeared, it does not reappear" (344), and we have seen in chapter 2 that Darwin thought that extinction is a law of nature. But we have to be extremely careful here. To say that in the geological record extinct groups do not reappear is not necessarily to say that it is logically impossible, or even physically impossible, that they could reappear. Indeed elsewhere in the *Origin* Darwin says "When a species has once disappeared from the face of the earth, we have reason to believe that the same identical form never reappears" (313), the reason being "for the link of generation has been broken" (344). Even if a lineage should come to occupy a niche left vacant by an extinct species, the new occupant of the niche is not the same species, "for both would almost certainly inherit different characters from their distinct progenitors" (315). Notice the key words in the first and third of these passages, namely, "we have reason to believe" and "almost certainly." Darwin is pretty clear here that he does think it could happen, that an extinct species could reappear, just that it is extremely unlikely.

Darwin is clearer on this in his correspondence. In a letter to Lyell (June 21, 1859) he wrote "You ask about specific centres, if you change terms into specific areas, my theory quite requires them; i.e. it is, I think, next door to an impossibility that the same species should have been formed identically the same in any two areas" (Burkhardt and Smith 1991, 307–308). "Next door to an impossibility," of course, is nevertheless to affirm a *logical possibility*. Thus for Darwin it is logically possible, albeit highly improbable, that someday, to quote from Lyell (1830), "The huge iguanodon might reappear in the woods, and the ichthyosaur in the sea, while the pterodactyle might flit again through umbrageous groves of tree ferns" (123).

Darwin held the same view on languages. In *Descent* (1871 I) he says "A language, like a species, when once extinct, never, as Sir C. Lyell remarks, reappears" (60). I suggest that the word "never" here should not be taken very strongly. As we have seen with the *Origin* and from Darwin's correspondence, he thought that the reappearance of an extinct species is highly unlikely. In the case of languages this interpretation is only strengthened if we turn to the passage in Lyell's writings which Darwin refers to in the *Descent*. Lyell's remark is to be found in *The Antiquity of Man* (1863), in which he says "a language which has once died out can never be revived, since the same assemblage of conditions can never be restored" (467). It may not be clear whether Lyell's claim here is a claim of physical necessity or merely only probability. However, it is surely significant that much earlier Lyell (1838) stated only as a simple matter of geological observation that no species "ever reappeared after once dying out" (275; cf. Darwin Notebook E, Barrett *et al.* 1987, 105). Of much greater significance is Lyell's (1830) postulation of a vast, geologically recurring "great year" (ch. 7), such that, as we have seen above, huge iguanodons, ichthyosaurs, and pterodactyles will once again populate the earth.

Given, then, that Darwin thought that the extinction of a species, like that of a language, is not necessarily forever, it follows that his concept of common descent was not a concept of monophyly in any of the forms that we encounter today (each of which insist on a single origin for each species). Common descent was not for Darwin a matter of logic or methodology, but rather only probability. The possibility that a particular species could arise independently in more than one place and time was so improbable to him that it was not worth worrying about.

This, of course, makes Darwin's species concept radically different from the species-as-individuals thesis of Ghiselin and Hull and from phylogenetic taxonomy, both of which hold that extinction is necessarily (logically) forever and that species taxa must each have no more than a single origin. Combined with what we have seen in the previous chapter, namely, Darwin's emphasis on the horizontal dimension for the reality of species, we can only conclude that common descent is not part of the ontology of species for Darwin but instead serves as, or rather the evidence for it serves as, an important criterion in picking out horizontal species (as well as higher groups). To use a modern distinction, the difference is between constitutive properties and diagnostic criteria (akin to the distinction between a disease and its symptoms), common descent being diagnostic but not constitutive of species.

It is also important to add here that although Darwin thought of common descent as a criterion of species delimitation, he did not think of it as the most important criterion. What was most important for him is what we shall examine in the next chapter. But for the remainder of the present chapter I want to focus more closely on Darwin's emphasis on common descent as well as the emphasis on monophyly in modern systematics.

Part of Darwin's argument was that naturalists were already using common descent in their classification of organisms into species. As he puts it in the concluding chapter of the *Origin*, "every naturalist has in fact brought descent into his classification; for he includes in his lowest grade, or that of a species, the two sexes; and how enormously these sometimes differ in the most important characters, is known to every naturalist" (424). Indeed then as today, no naturalist, as Darwin says, "dreams of separating them." Not even extreme splitters would ever dream of this. And if the sexes were ever separated by a taxonomist, it was only by accident, such that once the mistake became known it was quickly corrected. Such accidents are caused by the fact that many species are remarkably sexually dimorphic. For example, in the case of woodpeckers the sexes have evolved niche differentiation to such a degree that there is an extreme dimorphism in bill morphology. In the case of fur seals, the male is several times larger than the female. And in the case of mallard ducks, the colors are so different that one would naturally be inclined to think that they are different species if one did not actually see them copulate (indeed, according to Mayr and Short 1970, 88, Linnaeus himself made the mistake of classifying male and female mallards into separate species).

Darwin goes on to point out that naturalists also use descent for keeping larval and adult stages in the same species. As well, he says, the naturalist "includes monsters; he includes varieties, not solely because they resemble the parent-form, but because they are descended from it. He who believes that the cowslip is descended from the primrose, or conversely, ranks them together as a single species, and gives a single definition" (424).

Mayr (1957a) surmises that pre-Darwinian taxonomists emphasized common descent below the species level because as essentialists it was the only way they could deal with variation:

> What is unexpected for this pre-Darwinian period . . . is the frequency
> with which "common descent" is included in species definitions. When
> such an emphatically anti-evolutionary author as v. Baer (1828) defines the
> species as "the sum of the individuals that are united by common descent,"
> it becomes evident that he does not refer to evolution. . . . Expressions
> like "community of origin" or "individus descendants des parents com-
> muns" (Cuvier) are frequent in the literature. These are actually attempts
> at reconciling a typological species concept (with its stress of constancy)
> with the observed morphological variation. [7]

This is an interesting suggestion and I won't pursue it any further. What is important for our present purpose is that in the *Origin* Darwin goes on to point out that "may not this same element of descent have been unconsciously used in grouping species under genera, and genera under higher groups, though in these cases the modification has been greater in degree, and has taken a longer time to complete? I believe it has thus been unconsciously used; and only thus can I understand the several rules and guides which have been followed by our best systematists" (425). Darwin, however, was doing more than simply arguing that systematists, in classi- fying groups within groups, were unconsciously employing common de- scent. As we have seen at the beginning of this chapter, he also argued that it *ought* to be so for any natural system. Indeed we can see Darwin using a consistency argument. If you use common descent in classifying sexes, on- togenetic stages, sports, and varieties into one species (and you all do), and if branching evolution is true (and Darwin argued that it is), then you ought also to use common descent when grouping higher taxa.[2]

What we have to keep in mind in all of this, however, is that Darwin repeatedly claimed that common descent (he used the term "genealogy") was not enough for classification. The *Origin* itself caused some confusion about this and in his correspondence Darwin attempts to clear it up. For example, in a letter to Hooker (December 23, 1859) Darwin wrote "geneal- ogy by itself does not give classification" (Burkhardt and Smith 1991, 444), and in a letter to Andrew Murray (April 28, 1860) he wrote "In case of classification, descent alone, as I believe I have shown, will not do; you must combine principle of divergence of character & descent" (Burkhardt *et al.* 1993, 179).

So Darwin was certainly not a cladist, but so what? What is instruc- tive about Darwin's emphasis on common descent, or rather his lack of ex- clusive emphasis on common descent, is that it provides a much more sensible approach to the topic of multiple origins than that found today among cladists.

Let's see how this is. Darwin was arguably wrong in his claims against multiple origins (including that an extinct species almost certainly never reappears). What is now well-known is polyploidy, a chromosomal accident in a zygote resulting in at least three times the haploid (half) somatic chromosome number of the parent(s). When only one species is involved it is called *autopolyploidy*, when two species are involved (as in interspecific hybridization) it is called *allopolyploidy*. Although quite rare in animals, it is common in plants. The interesting thing about polyploidy is that it is normally considered *instantaneous speciation* by biologists, since the polyploids are reproductively isolated from their parental species. Indeed so common is polyploidy in plants that it is estimated that 70–80% of all angiosperm (flowering plant) species in the wild were produced by polyploidy. We therefore don't need science fiction examples (or genetic engineering) to discuss realistically the related issues of multiple origins and the reappearance of extinct species. In the previous chapter I cited the example of multiple origins given by Ashton and Abbott (1992), and there are many more (cf. Stamos 2003, 320 n. 15). The problem is that repeated polyploidy, if genealogy is taken strictly with similarity being completely irrelevant, arguably results in multiple species, not one species, *no matter how qualitively similar the resulting populations.* Strict cladists are forced to this conclusion, because for them only monophyletic taxa are natural taxa, so that a species cannot have multiple origins. Thus for them, if species A and species B repeatedly hybridize and produce 20 polyploid populations, *each* of those polyploid populations (no matter how genetically and morphologically identical, no matter how reproductively compatible) would have to be counted as a distinct species with a distinct binomial. What is worse, since each branching point is a speciation event for cladists—and only branching points count for cladists as speciation events, such that at any branching point (whether speciation by splitting or speciation by budding) the parental species automatically goes extinct (this has been dubbed "Hennigian extinction" by one of its critics)—both of the parental species automatically go extinct at the point of polyploid hybridization and become distinctly new species, even though they have not changed at all (cf. Ridley 1989, 5; Stamos 2003, 265)!

Had Darwin known about polyploidy, I think it is safe to say he would have thought the cladistic interpretation of it pure nonsense. It is as ridiculous for species as it would be for languages. He would see the cladistic concept of monophyly for what it is, *viz.* a convention, not a causal process, its mistake being that it confuses methodology with ontology (cf. Stamos 2003, 309). In the *Origin* Darwin writes of "the *vera causa* of community of descent" (159), but I think the case of polyploidy would have made him agree that common descent is only really a subsidiary *vera causa* of species ontology, it is not the main one (as we shall see in the next chapter). *Logically speaking*, Darwin had no objection to multiple origins. *If* species could arise multiply from the same parental species, then, just like languages if the same were to occur, he would rank all of the new productions as one. We shall see from the next chapter that Darwin would be forced to this conclusion because of his emphasis on adaptations. *If* the polyploids had new adaptations, then the multiple polyploids would be conspecific. Whether they were interfertile with each other and intersterile with their parental species would have nothing to do with it, as we shall see in chapter 6. Since, again as we shall see in the next chapter, Darwin thought it highly unlikely that new adaptations are produced by anything except gradual, cumulative natural selection, it is highly doubtful that he would give the polyploids new specific status.[3] At any rate, what Darwin's species concept makes us realize is that although common descent was very important for him as a species criterion, it was nevertheless not absolute and in some cases would not apply.

What the case of repeated polyploidy also relates to is the matter of whether the extinction of a species is necessarily forever. Suppose our species A and species B were to produce a polyploid population that subsequently went extinct, and that, a season later, species A and species B were to do it again. According to cladism, each of the two polyploid populations, the former and the latter, would have to be distinct species, no matter how qualitatively similar. Darwin, however, as we have seen in the previous paragraph and shall see from the next two chapters, would probably not agree that the first polyploid population is a new species, because it is a kind of hybrid and almost certainly would not have any new adaptations. So likewise in the case of repeated polyploidy, whether in the above scenario or in the previous scenario, he would not consider multiple origins multiple species. Nevertheless, *if* the new polyploid population that went extinct had a new adaptation, and *if* the second polyploid population was just like it, then Darwin would have no problem (as with many biologists today) classifying the two polyploid populations as conspecific. In other words, just as in the case of languages if the same were to occur, he would agree that an extinct species reappeared. This consequence, however, is logically precluded by cladistic principles and the species-as-individuals view.

Darwin, as we can see from the above, including his view on extinction, and as we shall see in the next chapter, was not against similarity in the ontology of species, contrary to cladists and proponents of species as individuals. And again, guided by the language analogy, this makes good sense. As we have seen in the previous chapter, according to the view of the above philosophies a species that undergoes "infinite evolution," to use the phrase of Ridley (1989, 10), is numerically the same species throughout. I don't think Darwin would have agreed with this, again, because he thought "genealogy by itself does not give classification" and he was not completely against similarity. In fact, many cladists are not completely against similarity either! I have found a significant number of cases where cladists, contrary to the principles of Willi Hennig, refuse to allow that a species that buds off a founder population (which in turn becomes a species), but that itself remains unchanged, automatically goes extinct at that branching point and becomes a new species (Stamos 2003, 262). In such cases, as I put it, these cladists break with cladism proper and "allow similarity to slip in through the back door" (263). Indeed arguably, if one wants to have a sensible ontology for species, one cannot avoid similarity. Of course one should not go to the other extreme either, involving only similarity, as with the morphological and phenetic species concepts. This is to invite a whole new set of problems and absurdities (cf. Stamos 2003, 81–83, 129–133, 311–312). Instead it would seem that the most viable position, and I think Darwin would agree as we shall see in the next chapter, is to find some sort of way of combining similarity with other considerations important in the ontology of species, which is what I attempted in my species book (Stamos 2003, ch. 5).

It is often argued that a species concept, to be viable, must be consistent with history reconstruction. For example, this was Donn Rosen's (1979) complaint against the biological species concept of Mayr. Rosen's paper is considered the inauguration of so-called phylogenetic species concepts, species concepts that attempt to make species consistent with phylogenetic history but that are not necessarily cladistic, and it is important to look more closely at his claims. According to Rosen, proponents of the biological species concept such as Mayr sometimes unite into a single species two good and distinct species that have a hybrid zone between them (276). The problem with this, for Rosen, is that reproductive compatibility is an ancestral character (*plesiomorphy*) that is gradually lost in the descendant species during geographic isolation. Moreover, he says, "It is to be expected that reproductive compatibility, like other primitive traits, might be retained or altered in a mosaic pattern during evolution, an inference which is entirely consistent with the results of natural and laboratory mating patterns in *Xiphophorus*" (277). The conclusion follows, then, that "the 'biological species concept' will lead to inferences that are in direct conflict with the avowed aims of systematics, viz., to reconstruct the genealogical history of lineages by a process of estimating a hierarchy of relationships" (277). Rosen then goes on to give the first version of what has later been called a phylogenetic species concept, in his case one that defines a species as having "one or more apomorphous features" (277) (an *apomorphy* is a derived character, not an ancestral character) and having "a specifiable geographic integrity" (278). Thus conceived, he says, "all populations [species by his definition] defined by apomorphic traits can be incorporated into a cladistic hierarchy, and . . . this cladistic hierarchy forms the only logical basis for discussions of the history of organic change in time and space" (277).

In my species book (Stamos 2003, 84–85, 273) I provide a critique of Rosen's phylogenetic species concept, but that is not what I want to bring to focus here. Instead, it is the claim that a species concept should be consistent with history reconstruction. The problem is that nature is not perfectly hierarchical as cladistics desires. As we have seen above, nature includes rampant polyphyly, and allopolyploidy is not the only example (cf. Stamos 2003, 240 n. 46). In focusing on the importance of genealogy for classification, Darwin was surely right, but he was also surely right in not focusing exclusively on genealogy. The case of multiple origins alone shows that similarity must also be given some account. But to do so does not necessarily preclude responsible history reconstruction. In fact, in some cases it might be necessary. The case of repeated polyploidy speciation discovered by Ashton and Abbott (1992) is a case of history reconstruction, but it was not done using cladistic principles; indeed this would have been impossible. As McDade (1995) put it, "current phylogenetic methods are inappropriate for taxa with complex reticulating histories, and yet the phylogenetic history of such groups is at least as interesting as those of divergently evolving groups" (616–617). At the species level especially, then, something more is needed than cladistic principles. As the example of Ashton and Abbott makes clear, good history reconstruction at the species level is not simply a matter of distinguishing the plesiomorphic from the apomorphic characters. Instead one has to use all the evidence available (cf. Stamos 2003, 318–320 for a detailed discussion on Ashton and Abbott 1992). And indeed if history reconstruction is really what one wants, if it is what really matters, then why confine oneself to only one methodology? This seems to me a misplaced loyalty.

Here again we can see the value of Darwin. In focusing on the horizontal dimension for the ontology of species, in analogy with the ontology of languages, he avoids in one stroke the absurdities involved with cladistic species concepts and the species-as-individuals view. In focusing on genealogy, however, species must be consistent with history reconstruction. But this, for Darwin, did not preclude multiple origins or the reappearance of an extinct species. In all of this, again, Darwin was guided very sensibly by his language analogy. In historical linguistics, languages are considered historical entities, and reconstructing language history is a thriving exercise (cf., e.g., Fox 1995; Trask 1996; Lass 1997), and yet languages are viewed as primarily horizontal entities, not as individuals, and language reconstruction, because of even greater reticulation in their histories, is only partially guided by cladistics (cf. Hoenigswald and Wiener 1987, especially the paper by Wiener).

Edward Wiley (1981, 74–75) has made the useful suggestion that the ontology of biology should admit of not two (as with Ghiselin and Hull) but of three categories, namely, class, individual, and historical group. In his view, the species category is a class, species are individuals, and higher taxa are historical groups (because even though they are spatiotemporally bounded they lack cohesion). What is needed, it seems to me, and Darwin would seem to be in agreement, is the recognition not only that species are not individuals, but that the category "historical group" should be changed to "historical entity." As primarily horizontal entities demarcated in part by common descent, species are historical entities, not individuals or classes. Individuals are historical entities too, but of a very different kind. As vertical entities they contain their history within themselves (or rather a large part of it). Historical entities such as species and languages, however, do not as horizontal entities contain their history within themselves. But they are nonetheless historical for it. As historical entities that are not individuals, that are not even groups, they are not spatiotemporally bounded but are free to have multiple origins and to go extinct and re-evolve, whether at the same time and in different places or at different times and places. But as dynamic entities they are not the timeless and fixed entities of abstract classes either. These categories (class and individual) just do not at all adequately capture the ontology of species. At an intuitive level, guided by the language analogy, Darwin knew this.

In sum, in conceiving of species as primarily horizontal, and yet in maintaining that all true classification must be genealogical, Darwin was not involving himself in a contradiction. In modern linguistics, particular languages are thought of as primarily horizontal (synchronic) entities, even though historical linguistics thrives. And this is in spite of getting over, roughly 100 years ago, the view that languages are individuals. Biology, however, seems to be over 100 years behind. In spite of agreeing that true classification must be genealogical, that species (like languages) are historical entities, many biologists (and many philosophers) seem to think that this means that species must be individuals. But it just doesn't follow. Darwin knew this. Post-Synthesis biologists such as Mayr knew this. During the last 30 or 40 years, however, mainly because of the influence of cladism and the species-as-individuals view, biology and its philosophy seem to have regressed. Fortunately, as pointed out at the end of the previous chapter, there does seem to be a trend within the field of phylogenetic systematics itself, back toward a primarily horizontal conception of species, while at the same time retaining the view that species are historical entities. This is a trend that Darwin would certainly have welcomed.

Chapter 5

Natural Selection
and the Unity of Science

In leading up to what is for Darwin the most important criterion in species delimitation, it will be useful to look at the main feature of John Beatty's theory, which as we shall see in chapter 8 has become the received view. Following Ghiselin's (1969) view that Darwin was a species taxa though not species category realist, so that Darwin's nominalistic definitions of "species" applied only to the species category, Beatty (1985) provided a strategy theory to explain Darwin's nominalistic definitions of "species." According to Beatty, Darwin distinguished

> between what his fellow naturalists *called* "species" and the non-evolutionary beliefs in terms of which they *defined* "species." Regardless of their definitions, he argued, what they *called* "species" evolved. His species concept was therefore interestingly minimal: species were, for Darwin, just what expert naturalists *called* "species." By trying to talk about the same things that his contemporaries were talking about, he hoped that his language would conform satisfactorily enough for him to communicate his position to them. [266]

Again and again, we are told by Beatty that Darwin "tried to get beyond definitions to referents" (269), that "The evolution issue was accordingly, for Darwin, an issue concerning the *species so designated by naturalists*" (274),

that "he used the term ['species'] in accordance with *examples* of its referential use by members of his naturalist community" (277), and moreover
that this "also allowed Darwin to communicate the position that the term
'species' was undefinable" (274).

I shall look more closely at the various aspects of Beatty's strategy theory in chapters 8 and 9. For the present, I want to focus on his claim that
Darwin simply followed the species designations of his fellow naturalists.
The other side of the coin to this claim is that Darwin did not try to change
any of the species designations of his fellow naturalists. Both claims, of
course, are factual claims, and accordingly they are to be decided by nothing else but an appeal to the extant evidence. When we look at that evidence, however, we cannot help but be struck by the fact that Darwin often
went *against* what his fellow naturalists called "species," and often decided
one way or another in cases where they did not agree. This is a striking fact,
which creates a serious problem for anyone who claims that Darwin was a
species nominalist, taxa or category. But even more important is what is to
be learned when we look closely at those cases where Darwin went against
his fellow naturalists.

A good place to begin is with a close look at what Darwin wrote in the
Origin about primroses and cowslips. But first, it is important to notice a
remark made by Malcolm Kottler (1978), that "Both Darwin and Wallace
were fully aware of well-defined species. But they were more interested in
borderline cases . . . *not* for the purpose of proving that species did not really exist, but rather for the purpose of disproving special creation. If species
had been independently created, borderline cases should not exist. Yet such
doubtful forms were abundant" (297). This claim about the importance
of borderline cases, of course, is surely true. It was necessary for Darwin
that the existence of borderline cases was proved to exist so that varieties
could be claimed to be incipient species in accordance with his thesis of
gradualism. As Darwin put it in the *Origin*, "wherever many closely-allied
species occur, there will be found many forms which some naturalists rank
as distinct species, and some as varieties; these doubtful forms showing us
the steps in the process of modification" (404).

But even more important for Darwin's argument against special creation, more important than borderline cases between varieties and species,
were cases that according to the general species concept of creationists
would have to be classified as *both* varieties and species, in other words,
cases that proved that the traditional criteria were incoherent and that therefore served as a falsification of the creationist species concept.

Such was the case of primroses and cowslips, and it explains why Darwin was so fond of referring to them. The problem is this: As Darwin tells us in the *Origin*, primroses and cowslips are "united by many intermediate links" (50) and so are "generally acknowledged to be merely varieties" (485). In other words, according to the traditional, creationist view, "they grant some little variability to each species, but when they meet with a somewhat greater amount of difference between any two forms, they rank both as species, unless they are enabled to connect them together by close intermediate gradations" (297). Thus, part of the creationist species concept involved the criterion that if any two forms are connected by intermediate gradations, no matter how different they may be, they must be classified as conspecific varieties rather than as separate species. But another important part of the creationist species concept was the sterility criterion. As Darwin put it, "The view generally entertained by naturalists is that species, when intercrossed, have been specially endowed with the quality of sterility, in order to prevent the confusion of all organic forms" (245). And again, as regards varieties, "a supposed variety if infertile [intersterile] in any degree would generally be ranked as species" (271).

This is what makes primroses and cowslips so interesting. United by many intermediate links and therefore ranked as varieties, Darwin on the other hand also informs us that "according to very numerous experiments made during several years by that most careful observer Gärtner, they can be crossed only with much difficulty" (49–50), so that Gärtner "ranks them as undoubted species" (268). Thus, according to creationist criteria, primroses and cowslips must be ranked as *both* varieties and species.

Of even more interest and importance than the negative consequences of primroses and cowslips for the creationist species concept, however, is what Darwin's classification of them reveals to us about his own species concept. For, unlike most other naturalists, and therefore contrary to Beatty's strategy theory, Darwin classified primroses and cowslips as separate species. As he put it in the *Origin*, he thought them "worthy of specific names" and that "in this case scientific and common language will come into accordance" (485).

At first sight, this is extremely odd. As Kottler (1978) points out, "In the *Origin* . . . the only explicit distinction which he [Darwin] drew between variety and species was the presence or absence of intermediate gradations, *not* the presence or absence of interbreeding" (297). As Darwin in the *Origin* put it, "Hereafter we shall be compelled to acknowledge that the only distinction between species and well-marked varieties is, that the

latter are known, or believed, to be connected at the present day by intermediate gradations, whereas species were formerly thus connected" (485).
This is, of course, and significantly for my analysis in chapter 3, a purely
horizontal criterion. And notably it is followed by Darwin in *Descent* (1871
I), where he takes sides on the then prevailing controversy over whether
the different human races are conspecific varieties (monogenism) or con-
generic species (polygenism), a debate that had an interesting relation to the
debate over human slavery (cf. Gould 1981, 69–72). Therein Darwin reaf-
firms that "Independently of blending from intercrossing, the complete absence, in a well-investigated region, of varieties linking together any two
closely-allied forms, is probably the most important of all the criterions of
their specific distinctness" (215). Accordingly "the most weighty of all the
arguments against treating the races of man as distinct species, is that they
graduate into each other, independently in many cases, as far as we can
judge, of their having intercrossed." To this Darwin adds that "Every naturalist who has had the misfortune to undertake the description of a group
of highly varying organisms, has encountered cases (I speak after experience) precisely like that of man; and if of a cautious disposition, he will
end up by uniting all the forms which graduate into each other as a single
species; for he will say to himself that he has no right to give names to objects which he cannot define" (226–227). Darwin's own experience, of
course, relates mostly back to his work on barnacles. And indeed in his Introduction to his work on the Balanidae (Darwin 1854b), he says "In determining what forms to call varieties, I have followed one common rule:
namely, the discovery of such closely allied, intermediate forms, that the application of a specific name to any one step in the series, was obviously impossible; or, when such intermediate forms have not actually been found,
the knowledge that the differences of structure in question were such as, in
several allied forms, certainly arose from variation" (156).

What is even more problematic is that in both cases, namely, that of
primroses and cowslips and the races of man, Darwin thought the evidence
preponderant that they each originated from a single stock. In the *Origin*
Darwin tells us that primroses and cowslips "are united by many intermediate links, and it is very doubtful whether these links are hybrids; and there is,
it seems to me, an overwhelming amount of experimental evidence, showing
that they descend from common parents" (50). Similarly in *Descent* (1871 I)
Darwin tells us that "Those naturalists . . . who admit the principle of evolution . . . will feel no doubt that all the races of man are descended from a
single primitive stock; whether or not they think fit to designate them as distinct species, for the sake of expressing their amount of difference" (229).[1]

To anyone who has read Darwin's chapter on classification in the *Origin*, Chapter XIII, the phrase "amount of difference" should ring a bell. Although Darwin tells us therein that "all true classification is genealogical" and that classification "must be strictly genealogical in order to be natural," he immediately adds that "the *amount* of difference in the several branches or groups . . . *is* expressed by the forms being ranked under different genera, families, sections, or orders (420, second italics added).

Kevin Padian (1999) argues that "Darwin chose his verbs carefully" (357), that the *is* refers not to Darwin's own views on the matter but only to the practices of his fellow naturalists, and that Darwin himself did not champion dual criteria for classification (genealogy and similarity) but only one criterion (genealogy), so that for Darwin "similarity is not in itself a criterion for classification, but [merely] a means to understand genealogy" (356). Padian also focuses on Darwin's letter to Hooker (December 23, 1859) that I quoted in the previous chapter, in which Darwin wrote "genealogy by itself does not give classification" (Burkhardt and Smith 1991, 444), and argues that the context of the letter, which is Darwin's response to the evolutionary views of the French botanist C.V. Naudin, indicate that Darwin had misunderstood Naudin, such that "Darwin did not mean *should* when he said 'genealogy alone *does* not give classification.' In other words, he meant that in practice it does not, and he could not see how Naudin (as he thought) could say that it did" (360).

Padian's claims, however, do not stand up to close scrutiny, and all the evidence does indeed strongly point to the conclusion that Darwin thought that similarity should be taken to be *constitutive* of taxa (species in particular), though not exclusively similarity. To return to the passage in the *Origin* that Padian reinterprets, the *is* does indeed refer to Darwin's own views, and Darwin makes this clear only a couple of pages further, where he states "Thus, on the view which I hold, the natural system is genealogical in its arrangement, like a pedigree; but the *degrees of modification* which the different groups have undergone, *have to be* expressed by ranking them under different so-called genera, sub-families, families, sections, orders, and classes" (422, italics added). Moreover, to Padian's highly questionable reinterpretation of Darwin's letter to Hooker above, we need only add Darwin's letter to Andrew Murray (April 28, 1860), in which he wrote "In case of classification, descent alone, as I believe I have shown, will not do; you must combine principle of divergence of character & descent" (Burkhardt *et al.* 1993, 179). Finally, what clinches the matter, especially with regard to species taxa, is Darwin's view on extinction, which as we have seen in the previous chapter did not preclude the possibility that an extinct species could reappear.

Returning to the issue of "amount of difference" in the *Origin*, we can see the role it plays for Darwin in determining whether two forms connected by intermediate gradations ought to be classified as varieties or species, when he says "Hence, without quite rejecting the consideration of the present existence of intermediate gradations between any two forms, we shall be led to weigh more carefully and to value higher the actual amount of difference between them" (485). The example he then gives is that of primroses and cowslips! Similarly in *Descent* (1871 I) Darwin tells us that "Nevertheless it must be confessed that there are forms, at least in the vegetable kingdom, which we cannot avoid naming as species, but which are connected together, independently of intercrossing, by numberless gradations" (227). In the footnote to this passage (n. 18), he refers to examples given by Nägeli, also Gray, but without referring to any species names. Interestingly, in a letter to Lyell (May 8, 1860) Darwin wrote "With respect to Aster, I remember long ago reading curious paper by Asa Gray & another on Aster, in which they give cases of two forms so very distinct that they must consider them specifically distinct, & yet perfectly united by intermediate varieties or links & they admit that in almost every other case that this suffices to upset two species as distinct" (Burkhardt *et al.* 1993, 196).[2]

What "amount of difference" in Darwin's theory of classification largely signifies, then, especially for his species concept, is that similarity relations must be viewed as *constitutive* of taxa. Again, this conclusion is entailed alone by Darwin's view that extinction is not necessarily (logically) forever. Nevertheless this is not to say that Darwin thought of species in terms of overall similarity. Scholars such as Sulloway (1979, 43), Mayr (1982, 210, 1991, 30), and Ruse (1987, 344) have interpreted Darwin this way, but such a view cannot possibly hold up. Granted, in the *Origin* Darwin says "I look at the term species, as one arbitrarily given for the sake of convenience to a set of individuals closely resembling each other" (52). But this is Darwin speaking at his loosest, and possibly also strategically (cf. chapter 8). When we look more closely, at his chapter on classification, for example, we have seen in the previous chapter that he disassociates himself from the view that classification is "placing together the forms most like each other" (414), "the mere putting together and separating objects more or less alike" (420). At the species level this was especially evident in his work on barnacles. When Darwin discovered what he came to call "complemental males," males so small in relation to the hermaphrodites to which they were attached (such that naturalists had thought they were parasites of the latter), he nevertheless thought of them, in spite of their being "utterly different in appearance and structure" from the hermaphrodites, as "belonging to the same species!" (Darwin 1851b, 293). In the *Origin* the same conclusion is maintained, and as we have seen in the previous chapter other kinds of examples are added to illustrate the dominance of common descent over similarity, such as larval and adult stages of the same individual (424). In addition, we have seen in the previous chapter what Darwin thought of convergence, even at the species level. So although Darwin included similarity as constitutive of species, his concept was far from that of overall similarity. Common descent was one of the relations that put bounds on the similarity complexes. Another was the primacy of the horizontal dimension, which, along with common descent, intergrading, and amount of difference, helped to distinguish species from varieties.

Nevertheless, if the only difference between primroses and cowslips being two species and the races of man being one species is amount of difference, then amount of difference sounds pretty arbitrary. And if that's what Darwin's species concept comes down to, then it's not much of an objective species concept. This is a valid criticism. But it turns out that there is much more involved in Darwin's species concept than mere amount of difference.

Part of what is involved is whether the amount of difference is relatively "constant and distinct." Early in the *Origin* Darwin wrote that "Those forms which possess in some considerable degree the character of species, but which are so closely similar to some other forms, or are so closely linked to them by intermediate gradations, that naturalists do not like to rank them as distinct species, are in several respects the most important for us. We have every reason to believe that many of these doubtful and closely-allied forms have permanently retained their characters in their own country for a long time; for as long, as far as we know, as have good and true species" (47). Similarly in his concluding chapter he tells us that as a consequence of his view "Systematists will only have to decide (not that this will be easy) whether any form be sufficiently constant and distinct from other forms, to be capable of definition; and if definable, whether the differences be sufficiently important to deserve a specific name" (484).

What makes the difference, then, what overrides both intermediate gradations and common descent in Darwin's species concept, is the further criterion of "constant and distinct" characters, characters as "permanent" (though not in the creationist sense) as the characters that demarcate "good and true" species. This is largely why primroses and cowslips, in Darwin's view, are two species instead of one. Indeed in Darwin's view in the *Origin* species of all kinds are only "well-marked and permanent varieties" (133), or to vary the expression "only strongly marked and fixed varieties"

(155). This, of course, given Darwin's evolutionism, in particular his gradualism, only makes sense in the horizontal dimension. From the vertical perspective, in Darwin's view, with the exception of living fossils there are no permanent or fixed characters.

Although an improvement over mere "amount of difference," characters "constant and distinct" might nevertheless also seem largely arbitrary, given Darwin's gradualism. In reply, however, it might be pointed out that in the passages from the *Origin* quoted in the second paragraph above Darwin is dealing with borderline cases in the horizontal dimension, what are often today called "messy situations." Although such borderline cases are entailed by Darwin's view of evolution by natural selection, we saw in chapter 3 that Darwin believed that in most situations in the horizontal dimension species are "tolerably well-defined objects," that they "do not at any one period present an inextricable chaos of varying and intermediate links" (177). Thus the existence of "good and true" species (horizontal) in Darwin's view is no more arbitrary, unreal, or invalidated than dusk and dawn, for example, invalidate the reality of night and day.

The problem with this reply, however, is that in his correspondence Darwin cited the "constant and distinct" criterion as his main criterion for species delimitation. For example, in a letter to Hooker (July 30, 1856), on the topic of barnacles, Darwin wrote "I differ from him [Lyell] in thinking that those who believe that species are *not* fixed will multiply specific names: I know in my own case my most frequent source of doubt was whether others would not think this or that was a God-created Barnacle & surely deserved a name. Otherwise I sh^d. only have thought whether the amount of difference & permanence was sufficient to justify a name" (Burkhardt and Smith 1990, 194). Moreover, years later in a letter to Hooker (October 22, 1864) Darwin wrote "The power of remaining for a good long period constant, I look at as the essence of a species, combined with an appreciable amount of difference" (Burkhardt *et al.* 2001, 376). Citing this last passage, Poulton (1903) claimed that "Darwin regarded persistence of form as an important criterion of a species" (91).

And yet in spite of this passage in Darwin's letter, as well as what we have seen in the *Origin*, there is still something more to Darwin's "amount of difference" than just relatively constant and distinct characters. Indeed to think otherwise is to miss the key that unlocks a full understanding of Darwin's species concept. In short, it turns out that the crucial difference, for Darwin, is whether those characters have been established by natural selection; in other words, whether they are *adaptations*. In the case of primroses and cowslips, Darwin reveals early in the *Origin* his reasons for going against his fellow naturalists and ranking them as two distinct species instead of one: "These plants differ considerably in appearance; they have a different flavour and emit a different odour; they flower at slightly different periods; they grow in somewhat different stations; they ascend mountains to different heights; they have different geographical ranges; and lastly, according to very numerous experiments made during several years by that most careful observer Gärtner, they can be crossed only with much difficulty. We could hardly wish for better evidence of the two forms being specifically distinct" (49–50). The criterion that Darwin lists last here, as we shall see in the next chapter, is a red herring, and can only have been thrown in by Darwin for persuasive effect. Darwin's real reasons are the ones he lists before it. Although Darwin does not come out and say it, his conclusion can only be, given that his book is about the *Origin of Species by Means of Natural Selection*, that the characters which distinguish primroses and cowslips as two species in his view are characters that have been established by means of natural selection. That he does not come out and say this is because in the *Origin* at that point he has not yet focused on natural selection; that focus comes two chapters later.

In the case of the races of man, on the other hand, Darwin in *Descent* (1871 I) is more explicit. In the penultimate passage he tell us that

> as far as we are enabled to judge (although always liable to error on this head) not one of the external differences between the races of man are of any direct or special service to him. . . . The variability of all the characteristic differences between the races, before referred to, likewise indicates that these differences cannot be of much importance; for, had they been important, they would long ago have been either fixed and preserved, or eliminated. In this respect man resembles those forms, called by naturalists protean or polymorphic, which have remained extremely variable, owing, as it seems, to their variations being of an indifferent nature, and consequently to their having escaped the action of natural selection. [248–249]

This is the key. We can see it in the title of Darwin's most famous book, *On the Origin of Species by Means of Natural Selection*. Moreover in the *Origin* Darwin tells us what the connection is between "constant and distinct characters" and natural selection, the connection being that natural selection is primarily responsible for the fixed state. Beginning (as Darwin typically does) with the evidence from domesticated animals, he says "if any part, or the whole animal, be neglected and no selection be applied, that part (for instance, the comb in the Dorking fowl) or the whole breed will cease to have a nearly uniform character. The breed will then be said to have degenerated" (152). Looking to nature, he then goes on to say "In rudimentary organs, and in those which have been but little specialized for any particular purpose, and perhaps in polymorphic groups, we see a nearly parallel natural case; for in such cases Natural Selection either has not or cannot have come into full play, and thus the organisation is left in a fluctuating condition" (152). Again he says "That the struggle between Natural Selection on the one hand, and the tendency to reversion and variability on the other hand, will in the course of time cease; and that the most abnormally developed organs may be made constant, I can see no reason to doubt" (153–154). Granted, though Darwin then goes on in the next few pages to argue that "specific characters are more variable than generic" (154), the point remains for him that species are "only strongly marked and fixed varieties" (155). This, moreover, will prove significant when, in chapter 7, we examine Darwin's view on the nature of varieties and how they differ from species.

As the key to Darwin's species concept, we can now also see why he would give such high though not exclusive importance to common descent in his species concept, as we have seen in chapter 4. Common descent means common inheritance, and it is by common inheritance that adaptations get shared. Conversely, it was for Darwin highly unlikely that the same adaptation could arise separately by saltation, as we shall see a little later in this chapter, or that adaptations could evolve separately in different lines of descent, as we have seen in the previous chapter on the topic of convergence at the species level and as we shall see again in the next chapter when we look at Darwin on niches.[3]

As the key we can also see why Darwin would think of species primarily though not exclusively as horizontal rather than vertical entities, as we have seen in chapter 3. Species ontologically for Darwin could not possibly be lineages or clades because adaptations come and go over time in a single lineage or clade. When remaining relatively unchanged, however, their reality could be extended over long periods of time, as with living fossils or species F in Darwin's diagram in the *Origin*.

Returning to the matter of examples, so far we have looked at the cases of primroses and cowslips and the races of man. There are further cases that not only go against Beatty's claims but further support the interpretation developed above. A prime example is the case of domestic pigeons. In a letter to Huxley (December 13, 1859) Darwin supplied an enclosure he had written for Huxley on pigeon breeding. In that enclosure he included some quotations from the writings of John Matthews Eaton. Although Eaton preferred to refer to domestic pigeons as "species," Darwin in his enclosure referred to the "Breeds" of pigeons, remarking that if Huxley could see the drawings which Darwin has he "would have grand display of extremes of diversity" (Burkhardt and Smith 1991, 428). Why would Darwin group all the breeds of pigeons together as one species (as he did in the *Origin*, 20–28), contrary to the breeder Eaton, if he thought that the designation of species and varieties was arbitrary? This raises an interesting question. As we have seen earlier in this chapter, Darwin ranked primroses and cowslips as two species instead of one, even though they were derived from a common stock. In the case of domestic pigeons, with their great diversity, Darwin tells us in the *Origin* "Altogether at least a score of pigeons might be chosen, which if shown to an ornithologist, and he were told that they were wild birds, would certainly, I think, be ranked by him as well-defined species. Moreover, I do not believe that any ornithologist would place the English carrier, the short-faced tumbler, the runt, the barb, pouter, and fantail in the same genus" (22–23). Darwin also believed that all the pigeon breeds derived from a common stock, the rock pigeon, *Columba livia* (23), and yet, unlike primroses and cowslips, and in spite of their great diversity, he did not rank any of the breeds of domestic pigeons as specifically distinct. So if common descent and being well-defined are not sufficient criteria for a species distinction, and they clearly are not in this case, then what for Darwin is? Again the answer appears to fit quite perfectly with my theory above. Contrary to the characters that functionally distinguish primroses and cowslips, which as we saw Darwin thought were adaptations produced by natural selection, the characters that distinguish pigeon breeds were produced by artificial selection, the product of mere human fancy, and thus were functionally superficial. As Darwin put it in *Variation* (1868 I) at the end of his two chapters on pigeons, "It is not likely that characters selected by the caprice of man should resemble differences preserved under natural conditions either from being of direct service to each species, or from standing in correlation with other modified and serviceable structures" (233).

This connects with what Darwin wrote in the *Origin*. Although there he does indeed state that varieties or races produced by artificial selection "show *adaptation* in their structure or in their habits to man's wants or fancies" (38, italics mine), he nevertheless adds that artificial selection focuses only on "external and visible characters," whereas natural selection acts "on every shade of constitutional difference" (83). Furthermore, "Man selects only for his own good; Nature only for that of the being which she tends." Finally, the wishes and efforts of man are "fleeting," his time is "short," so that "how poor will his products be, compared with those accumulated by nature during whole geological periods" (84). Thus, for Darwin, "nature's productions should be far 'truer' in character than man's productions" and "should be infinitely better adapted to the most complex conditions of life" (84). Unlike domestic productions, then, which are produced by artificial selection and accordingly are superficial, and furthermore are produced not for the good of the organism but only for mere human fancy, natural selection produces genuine adaptations and consequently real species.

Returning to organisms in a state of nature, there are yet further examples that go against Beatty but that support the above view. For example, in the case of primroses, George Bentham, in his *Handbook of British Flora*, published in 1858, made *Primula scotica* a variety of *Primula farinosa*, but Darwin seemed to disagree. As he wrote in a letter to Daniel Oliver (February 20, 1863), "would it not be worth while to tell him of Mr Scotts observation; for there can be no doubt that this difference indicates an important functional difference" (Burkhardt *et al.* 1999, 159). Indeed in his book on cross-fertilization Darwin (1876b) followed Lindley and treated *P. scotica* as a species (362).

Further examples come from Darwin's book on orchids (1862a). Right from the start, Darwin states "The object of the following work is to show that the contrivances by which orchids are fertilized, are as varied and almost as perfect as any of the most beautiful adaptations in the animal kingdom" (1). Moreover in the above letter to Oliver, Darwin wrote "Treviranus in his Review of the Orchids does not seem to appreciate at all the prettiness of the adaptations, which seems to me the cream of the case" (159). Adaptations also appear in that work as the key for specific identity in orchids, which sometimes set Darwin against the species designations of his fellow naturalists. A good example is the case of *Habenaria chlorantha*, which Bentham ranked as a variety of *H. bifolia*. In a letter to Bentham (June 22, 1861), Darwin wrote "I must think that you are mistaken in ranking Hab. chlorantha as a var. of H. bifolia: the pollen-masses & stigma differ more than in most of the best species of Orchis" (Burkhardt *et al.* 1994, 185). And indeed in his orchid book (1862a) Darwin explicitly calls these characters, which he used to treat these forms as distinct species, "cases of adaptation" (44, cf. 69–73).

Another orchid example is *Ophrys arachnites*, which Darwin tells us "is considered by some botanists as only a variety of the Bee Ophrys [*O. apifera*], by others as a distinct species" (51). He then goes on to point out the structural differences, stating in conclusion that "This plant, therefore, differs in every important respect from *O. apifera*, and seems to be much more closely allied to *O. aranifera*" (52), the implication being that Darwin himself considers *O. arachnites* a species because of its unique adaptations.

A related example is given in Darwin's description of *Ophrys apifera*, namely, that in "North Italy *Ophrys apifera, aranifera, arachnites,* and *scolopax* are connected by so many and such close intermediate links, that all seem to form a single species in accordance with the belief of Linnaeus, who grouped them all together under the name of *Ophrys insectifera*" (58–59). Darwin then goes on to implicitly split them. First, he says "The case therefore is an interesting one, as here we have forms which may be and generally have been ranked as true species, but which in North Italy have not as yet been fully differentiated." He then adds "Whether we rank the several forms of Ophrys as closely allied species or as mere varieties of the same species, it is remarkable that they should differ in a character of such physiological importance as the flowers of some being plainly adapted for self-fertilization, whilst the flowers of others are strictly adapted for cross-fertilization, being utterly sterile if not visited by insects" (59).

Further examples come from cases of sudden origin (saltation), which some naturalists took to be true species but which Darwin in every case refused to accept. In chapter 3 (note 2) we have seen Darwin's response to William Harvey's case of *Begonia frigida*, which Darwin recognized as a distinct form but never once called a "species."

Of equal if not greater relevance are two cases discussed in *Variation* (1868 I). The first is that of the Himalayan breed of rabbit, a distinct white breed with red eyes and brownish-black ears, nose, feet, and upper side of the tail, which, it was found, "notwithstanding their sudden origin, if kept separate, bred perfectly true," so that "Some good observers . . . stoutly maintained that they formed a new species" (113), giving them the name *Lepus nigripes*. Darwin, however, rejected their specific status, dismissing them as a case of "reversion" to the "not improbable" originators of their silver-grey parental species, namely, a supposed cross between "a black and albino variety" (115).

The second case from *Variation* is that of the "japanned" or "black-shouldered" form of peacock (peafowl), which appeared independently in Great Britain from common peafowl in at least seven well-authenticated cases (306–307), which "propagate their kind quite truly" (305), and which one "high authority," Philip Sclater, named a "distinct species, viz. *Pavo nigripennis*" (305). Darwin, however, granting that he has "heard of no other such case in the animal or vegetable kingdom," argues that the form is a "strongly marked variety or 'sport,' which tends at all times and in many places to reappear" (307). Interestingly, a number of years earlier, in a letter to Sclater (May 14, 1862), Darwin wrote "With four cases now recorded I would wager the P. nigripennis will prove a variety,—hardly more surprising in its origin, than the so-called Himalayan rabbit" (Burkhardt *et al.* 1997, 193).

The above three cases provide an interesting set of examples since, in arguing against the species ascriptions of his fellow naturalists, Darwin is not appealing to adaptations, as in the cases of primroses, cowslips, and orchids, but to the lack of new adaptations, as in the cases of humans and pigeons. But unlike the latter cases, the origins of the forms in the previous three paragraphs were by saltation. And we know from the *Origin* that Darwin thought it highly unlikely that an adaptation would ever result by saltation. Monstrosities, he said, which he took to mean "some considerable deviation of structure in one part," are either, he continued, "injurious to or not useful to the species" and are "not generally propagated" (44). Instead for Darwin it was slow, gradual, cumulative natural selection that produced adaptations, not saltations. Again as he put it in the *Origin*, this time appealing to the gradualism and uniformitarianism of the "modern geology" of Lyell, "so will natural selection, if it be a true principle, banish the belief of the continued creation of new organic beings, or of any great and sudden modification of their structure" (95–96; cf. Stauffer 1975, 223). Indeed Darwin just could not see adaptations arising by saltation. As he put it in a letter to Gray (August 11, 1860), "Please tell him [Prof. Parsons] that I reflected much on chance of favourable monstrosities (i.e. great & sudden variations) arising. I have, of course, no objection to them; indeed it w^d. be great aid; but I did not allude to subject [in the *Origin*], for after much labour I could find nothing which satisfied me of the probability of such occurrences. There seems to me in almost every case too much, too complex, & too beautiful adaptation in every structure to believe in its sudden production" (Burkhardt *et al.* 1993, 317). Indeed to think otherwise, for Darwin, smacked of miracles and creationism, not formation by natural law.

In sum, all of the above examples should be sufficient to put to rest the view that Darwin simply followed what his fellow naturalists called "species." In the next chapter I shall give an explicit reconstruction of Darwin's species concept, in the form of a definition, after examining what he did *not* include in it. For the remainder of this chapter I want to focus on Darwin's species concept, characterized as we have seen mainly by natural selection and adaptation, as fulfilling the *vera causa* (true cause) ideal of the philosophy of science of his day.

Darwin's world, as is well known, the world of the science of his day, was a world characterized by discovering laws of nature. Darwin did not reject this world but fully embraced it. Indeed Darwin himself defined nature in terms of laws. For example, in *Natural Selection* (Stauffer 1975) Darwin wrote "By nature, I mean the laws ordained by God to govern the Universe" (224). In *Variation* (1868 I) Darwin wrote, with the above deism now noticeably missing, "I mean by nature only the aggregate action and product of many natural laws" (7). Accordingly Darwin also defined science in terms of laws. In his autobiography (1876a) he wrote "science consists in grouping facts so that general laws or conclusions may be drawn from them" (70).

Although Darwin would have simply imbibed this emphasis on laws and the discovery of them by science in his early education, there were two books in particular, and to a lesser extent a third, which greatly influenced him. The third would be Lyell's *Principles of Geology*, in particular the first and third volumes, both of which were read by Darwin during his *Beagle* voyage. In those volumes Lyell repeatedly appealed to laws of geology in his argument for the gradual transformation of geological landscapes and seas, including fossils, all of which appealed to processes operating in the here and now. In the first volume (1830), for example, Lyell states right at the outset that "Geology . . . enquires into the causes of these changes," the changes being those that have taken place both in the organic and inorganic worlds, and that it aims at discovering "the laws *now* governing its animate and inanimate productions" (1). In the third volume (1833) Lyell gives a specific example, that of "the laws of earthquakes" which throw light on the "origin of mountains" (5). Interestingly Lyell claimed that geology should be restricted "in the first instance, to known causes, and then speculating on those which may be in activity in regions inaccessible to us" (3).

But by far the greatest influence on Darwin for the topic of laws was John Herschel's *A Preliminary Discourse on the Study of Natural History* (1830). Darwin began reading this book shortly after finishing his B.A. at Cambridge in late January 1831, almost a year before beginning his *Beagle* voyage (begun on December 31, 1831, and finished on October 2, 1836). Darwin read the book again, according to his reading notebooks, in 1838 (Burkhardt and Smith 1988, 456), very shortly after he first read Malthus (September 28 to October 3), which of course helped generate his earliest formulation of natural selection (Barrett *et al.* 1987, 375–376). In his autobiography (1876a) Darwin wrote that Herschel's book, along with Humboldt's *Personal Narrative of Travels*, "stirred up in me a burning zeal to add even the most humble contribution to the noble structure of natural science. No one or a dozen other books influenced me nearly so much as these two" (68).[4]

Herschel's book was quickly recognized by his contemporaries as the authoritative treatise on scientific method. What is especially important for our purposes is what Herschel wrote about causes and laws. He argued that the aim of science is to find explanations in terms of "an immediate producing cause," and that "If that cannot be ascertained, the next is to *generalize* the phenomenon, and include it, with others analogous to it, in the expression of some law, in the hope that its consideration, in a more advanced state of knowledge, may lead to the discovery of an adequate proximate cause." Explicitly following Newton, Herschel applied the term "*veræ causæ*," by which he meant "causes recognized as having a real existence in nature, and not being mere hypotheses or figments of the mind" (144).

A further important influence on Darwin's concept of laws of nature came from William Whewell's *History of the Inductive Sciences* (1837), which from Darwin's reading notebooks we know he had read in 1838 (Burkhardt and Smith 1988, 456), having previously skimmed parts of it (Burkhardt and Smith 1986, 186), and which again (his full reading) was very shortly after he read Malthus. Darwin was possibly also influenced by Whewell's *Philosophy of the Inductive Sciences* (1840), which we know from his reading notebooks he had planned to read, noting apparently in 1841, after reading a review of it by Herschel, that "I see I must study" (Burkhardt and Smith 1988, 446), but we don't know if he even cracked the book open at all.

Whewell differed from Herschel in not requiring direct or even analogical evidence of *verae causae*, but instead emphasized a *consilience of inductions*. By this term he referred to the proof of a *vera causa* being obtained when it explains inductions (generalizations) from a wide variety of areas.

As Michael Ruse (1979, 176–180) points out, Darwin employed both approaches to *verae causae*, the direct and analogical conception of Herschel and Lyell and the consilience conception of Whewell. In the *Origin* the Herschelian approach is to be found in the first chapter, with Darwin's examination of variation under domestication and artificial breeding. This Darwin called "the best and safest clue" (4) for understanding the origin of species in nature. Darwin then goes on to make the case for natural selection in nature, and further for its production of varieties from chance variations in individuals and for species from varieties. In the remainder of the *Origin*, Darwin employed the Whewellian approach to *verae causae*, amassing evidence from a widely different group of facts (instincts, the fossil record, biogeography, homology, embryology, rudimentary organs, and non-evolutionary classifications), each of them an induction and collectively a consilience of inductions (cf. Ruse 1975; Hodge 1977; and Ruse 2000 vs. Hodge 2000 on Darwin's debt to Whewell). Darwin, however, did not explicitly bring out this feature of his argument in the *Origin*, glossing it over instead by stating that "this volume is one long argument" (459; cf. 5–6).

Ruse (1979, 235–236) points out that as time went on and Darwin's "best and safest clue" came more and more under attack, Darwin relied increasingly on the Whewellian conception of *verae causae*. In fact, however, Darwin began his repeated appeal to consilience slightly *before* the publication of the *Origin*, commencing with his letters to those who received advance copies, and then continuing this appeal for quite some time (cf. Burkhardt and Smith 1991, 369, 374, 403, 404, 422, 435, 449, 460, Burkhardt *et al.* 1993, 76, 84, 91, 123, 129, 137, 195, 266, Burkhardt *et al.* 1994, 52, 54, 96, 99, 108, 135, Burkhardt *et al.* 1997, 255, Burkhardt *et al.* 1999, 155). Perhaps the best example of these passages is Darwin's letter to Samuel Pickworth Woodward (March 6, 1860), for the way he uses both concepts of *vera causa*:

> The fair way to view the argument of my book, I think, is to look at Natural Selection as a mere hypothesis (though rendered in some degree probable by the analogy of method of production of domestic races; & by what we know of the struggle for existence) & then to judge whether the mere hypothesis explains a large body of facts in Geographical Distribution, Geological Succession, & more especially in Classification, Homology, Embryology, Rudimentary Organs The hypothesis to me does seem to explain several independent large classes of facts; & this being so, I view the hypothesis as a theory having a high degree of probability of truth. [Burkhardt *et al.* 1993, 123]

Darwin would later write, in a letter to Alexander Goodman Moore (March 8, 1861), comparing the matter to the belief in ether and the undulatory theory of light, "which is now universally admitted as a true theory," that similarly because evolution by natural selection explains many diverse classes of facts *"therefore* I believe in its truth" (Burkhardt *et al.* 1994, 50).[5]

Returning to the influence of Herschel, Silvan Schweber (1985) has suggested that there were many other aspects of Herschel's influence on Darwin besides the *vera causa* ideal, including that for Herschel good science involves the decompounding and resolution of complex phenomena into simpler ones, as in chemistry with elements, so that, as Schweber puts it, "I believe Darwin was sensitive to these remarks of Herschel when he pondered what constituted an 'element' in his theory of natural selection. Were individuals the 'elements,' or were varieties and species 'elements'?" (48). Beyond this tantalizing glimpse, however, apparently related to the issue of the units of selection, Schweber ventures no further. What I would like to add at this point is a further influence that Schweber might have glimpsed but alas has missed, along with Ruse and Hodge and everyone else, namely, the effect on Darwin's mature species concept of what Herschel wrote on the relation between *verae causae* and *natural classification.*

It will no doubt be helpful to begin with a quick look at Herschel's (1830) concept of laws of nature, since as we have seen he intimately connects it with the concept of *vera causa*. Herschel's concept will prove to illuminate Darwin's concept of laws as revealed in the examples examined in chapter 2 but especially with the example examined so far in this chapter, namely, natural selection.

Herschel had two great categories of laws that he subsumed under laws of nature and that he stated as follows: "We may therefore regard a law of nature either, 1st, as a general proposition, announcing, in abstract terms, a whole group of particular facts relating to the behavior of natural agents in proposed circumstances; or, 2dly, as a proposition announcing that a whole class of individuals agreeing in one character agree also in another" (100). The latter he called "the relation of *constant association*, inasmuch as it asserts that in whatever individual the one character is found, the other will invariably be found also" (101). The other category is much more important for our purpose (and also for Herschel), namely, "the consideration of a proximate, if not an ultimate, cause" (101). The classic example, of course, for Herschel as well as for others was Newton's law of gravity, which Herschel called both a "force" and a "real cause" (198). Causal laws for Herschel are the explanatory laws (144), and he conceived of these as being essentially conditional in nature, such that "if such a case arise, such a course shall be followed" (36). More importantly he conceived of these as being eternally fixed, applying to kinds that might not yet or that might not ever have any members, as with chemical elements and compounds, such that "no chemist can doubt that it is *already fixed* what they will do when the case does occur. They will obey certain laws, of which we know nothing at present, but which must *be* already fixed, or they could not be laws" (36).

Compare all of this now with Darwin's concept of laws, including especially natural selection. In his abstract of Macculloch's *Proofs and Illustrations of the Attributes of God* (Barrett *et al.* 1987, 632–641), written in late 1838, he complains about using the deity to explain structures in organisms: "The explanation of types of structure in classes—as resulting from the *will* of the deity, to create animals on certain plans.—is no explanation—*it has not the character of a physical law*, & is therefore utterly useless—it foretells nothing because we know nothing of the will of the Deity, how it acts & whether constant or inconstant like that of Man.—the cause given we know not the effect" (55). Here we see both the importance of explanation and prediction, themes that would recur throughout Darwin's writings where he mentions laws. For example, in his autobiography (1876a), in criticism of Herbert Spencer's laws, Darwin wrote "they do not seem to me to be of any strictly scientific use. They partake more of the nature of definitions than of laws of nature. They do not aid one in predicting what will happen in any particular case" (109). Similarly Paley's explanation of adaptations "fails, now that the law of Natural Selection has been discovered." Indeed Darwin immediately adds that "Everything in nature is the result of fixed laws" (87). We can also see the conditional nature in Darwin's concept of natural selection as law, in for instance his letter to Charles Bunbury (February 9, 1860), in which Darwin dealt explicitly with whether natural selection is a "vera causa," and in which he wrote "Natural selection seems to me in so far in itself not be quite hypothetical, in as much if there be variability & a struggle for life, I cannot see how it can fail to come into play to some extent" (Burkhardt *et al.* 1993, 76). He then goes on to emphasize the importance "in giving a rational, instead of theological explanation of many known facts" (77).

Darwin also subscribed to laws of association, as with what came to be known as the Darwin-Knight law, that "it is a general law of nature (utterly ignorant though we may be of the meaning of the law) that no organic being self-fertilises itself for an eternity of generations; but that a cross with another individual is occasionally—perhaps at very long intervals—indispensable" (Darwin 1859, 97; cf. Darwin 1858, 21). But what is most important for our purposes is whether he thought of natural selection as a force, so that the law of natural selection is a force law akin to Herschel's acceptance of Newtonian gravity as a force law. As we have seen in chapter 2, Darwin sometimes in the *Origin* calls natural selection a "power," while other times he calls it a "process." Darwin's repeated use of the word "power" might indicate that he thought of natural selection as a force analogous to gravity or magnetism, but in a letter to Gray (November 29, 1857) Darwin made it clear that he thought of natural selection, as it is often thought of today (e.g., Endler 1986, 29–33; Haynes 1987, 5), not as a force but as a process, in fact as the result of a number of processes: "I had not thought of your objection of my using the term 'natural Selection' as an agent; I use it much as a geologist does the word Denudation, for an agent, expressing the result of several combined actions" (Burkhardt and Smith

1990, 492). Darwin's use of the word "agent" is interesting in relation to Herschel's use, for although Herschel seemed to restrict the word "agent" to force laws, such as Newton's law of gravity, Darwin expanded the concept to apply to process laws as well, such as natural selection.[6]

Darwin, of course, argued that species are produced by laws, beginning early in his career as an evolutionist. In, for example, Notebook E (Barrett *et al.* 1987), he says "the laws of variation of races, may be important in understanding laws of specific change.—When the laws of change are known.——then primary forms may be speculated on, & laws of life,—the end of Natural History, will be approximated to" (54). Similarly in a letter to Hooker (July 30, 1856) he wrote "I am, also, surprised at his [Lyell's] thinking it immaterial whether species are absolute or not: whenever it is proved that they all species are produced by generation, by laws of change what good evidence we shall have of the gaps in formations. And what a science Natural History will be, when we are in our graves, when all the laws of change are thought one of the most important parts of Natural History" (Burkhardt and Smith 1990, 194).

Since species for Darwin are produced by laws, in particular natural selection (recall the title of the *Origin*), this connects to the issue of laws and classification, and again to Herschel, for in Herschel (1830) we find a very clear connection between *verae causae* and classification. In his fifth chapter devoted to the topic, titled "Of the Classification of Natural Objects and Phenomena, and of Nomenclature," Herschel tells us not only, as one would expect, that classification is an important part of science, but much more importantly he distinguishes between *natural* and *artificial* systems of classification. Although both classes are groups "bound together by general resemblances" (135), the most basic of which are species (chemical and biological), it is evident from Herschel's chapter that the main difference between a natural and an artificial system is that the former is based on *verae causae* whereas the latter is not. Although Herschel has very little to say about biological species—he clearly thinks of them in the same way as chemical species, as "distinguished from each other by essential differences" (136)—his concept of natural classification comes out quite clearly in the case of mineral species. Lamenting that "There is no science in which the evils resulting from a rage for nomenclature have been felt to such an extent as in mineralogy" (139), his complaint nevertheless is not so much that the names for minerals have not been fixed, that "there is scarcely one [mineral species] which has not four or five names in different books" (139). Instead his main complaint is that many mineralogists raise mineral species into genera so that their resulting mineral species are based on arbitrary rather than on nomic differences, in his words that "*species* too definite to admit of mistake, are actually rendered *generic*, and extended to whole groups, comprising objects agreeing in nothing but the arbitrary heads of classifications from which the most important natural relations are professedly and purposely rejected" (140).

This is the equivalent of the complaint against splitters in biology, as opposed to lumpers, a debate that raged in Darwin's time (and continues in our own).[7] In the case of mineral species, it is evident that the "most important relations" that Herschel thought the splitters overlook are the laws of crystallography. These laws play a repeated and prominent role throughout Herschel's *Discourse*. In his section on mineralogy he tells us that crystallography, with the aid of inventions such as the reflecting goniometer, "furnished the mineralogist with the means of reducing all the forms of which a mineral is susceptible under one general type, or primitive form" (292). Accordingly, each mineral species has a peculiar "cleavage," a character, he tells us, "in the highest degree geometrical, and affording, as might naturally be supposed, the strongest evidence of its necessary connection with the intimate constitution of the substance" (291). He further tells us that until such cleavages were discovered by crystallography, the main name being Abbé Haüy (240), "no mineralogist could give any correct account of the real distinction between one mineral and another" (291). In his section on crystallography (239–245) he provides a short history, leading up to the present state of the science, repeatedly referring to the "laws of crystallography," which he earlier referred to as "laws . . . which limit the forms assumed by natural substances . . . to precise geometrical figures, with fixed angles and proportions" (123). Nevertheless these laws are only laws of "constant association," to use his term examined above, so that, even though they allow a certain amount of novel prediction (240), they can only point to causal/mechanical laws that are not yet known (245). Even so, crystallography in Herschel's view developed to the point where it could legitimately be called a "separate and independent branch of science," since it applies not only to mineralogy but to other areas, notably chemistry and optics (293).

This lesson could hardly have been lost on Darwin. Not only had Darwin read Herschel's *Discourse* twice, and not only had Lyell emphasized the importance of crystallography for geology (Lyell 1830, 2, 1833, 66), but the history of mineralogy (including the importance of Haüy, the nature of cleavages, and the laws of crystallography) is repeated in even greater detail in Whewell's *History* (1837). Even during his student days at Cambridge, Darwin must have known about the laws of crystallography, either from his main mentor John Stevens Henslow, professor of botany but previously professor of mineralogy (indeed Henslow had taught Darwin's brother Erasmus mineralogy), or from Adam Sedgwick, professor of geology, from whom Darwin received a crash course in geology and with whom he went on a field trip just before his *Beagle* voyage. Indeed, Darwin had even brought with him on the *Beagle* a reflecting goniometer, though he was not expert enough at using it and wrote to Henslow for a book to help (Burkhardt and Smith 1985, 251, 252 n. 3, 352, 395; cf. 3 for an early reference to a "small ganiometer" by Erasmus so that "we shall be able to separate the different crystalls in your cab").

Darwin, of course, went on to become one of the greatest geologists of his time, although the public normally thinks of him only as a theoretical biologist. At any rate, the point that needs to be made is that Darwin, by appealing to the *vera causa* of natural selection in the origin of species, as well, as we have seen in the previous chapter, to other *verae causae* such as "the *vera causa* of community of descent," was attempting to do for biological classification what others had done for other sciences, namely, he was applying laws to distinguish real from unreal species.[8] Or more generally, by employing the *vera causa* ideal Darwin was attempting to bring biology, both classification and explanation of the origin of species, into the unity of science.

Indeed comparisons between Herschel on crystallography and Darwin on natural selection and classification abound. One further comparison is that Herschel, as we have seen above, argued that the laws of crystallography apply to other sciences such as chemistry and optics, not only to crystallography, while in chapter 3 we saw that Darwin applied the *vera causa* of natural selection to linguistics, not just to biology. Of course there are also profound differences. For one, Herschel was an essentialist with regard to species, whether chemical, mineral, or biological, whereas Darwin was not with regard to biological species. Moreover, genealogy means nothing for the delimitation of species in chemistry or mineralogy, whereas for Darwin, as we have seen in the previous chapter, it meant a lot indeed.[9] But to focus on such differences is to miss what is really important, to miss what Darwin was really trying to do. In this vein Darwin wrote in a letter to Lyell (August 21, 1861) that "in the case of species . . . their formation has hitherto been viewed as beyond law,—in fact this branch of science is still with most people under its theological phase of development" (Burkhardt *et al.* 1994, 238), which compares with what he wrote to Lyell a few weeks earlier (August 1), that "I must think that such views of Asa Gray & Herschel merely show that the subject in their minds is in Comte's theological stage of science" (226–227).

We can see, then, that Darwin connected one of his fundamental insights, that adaptations are produced by natural selection, with the reality of species (and also varieties, as we shall see in chapter 7). He could have easily titled his book *On the Origin of Adaptations by Means of Natural Selection*. But he didn't. He titled it *On the Origin of Species* because he really did believe in the objective existence of species. Darwin's focus, then, was not just on adaptations, but on the broader framework of evolution, and for that he needed not only adaptations but varieties and species. Character evolution was not simply his focus, and it could not be, because he also wanted to argue for and explain the branching evolution of species.

So Darwin connected the production of adaptations by natural selection with the reality of species. But he did more. By employing the *vera causa* law of natural selection to explain both the existence of adaptations and the reality of species, Darwin brought evolutionary biology (and ultimately the rest of biology) into the framework of the natural sciences, which were based on laws.

An interesting question remains, however, whether Darwin thought that natural selection, and only natural selection, produces adaptations. There is, of course, the passage at the very end of the Introduction to the *Origin* in which Darwin says "I am convinced that Natural Selection has been the main but not exclusive means of modification" (6). This was a claim that Darwin repeated throughout his correspondence. One might naturally think, then, that Darwin was not a real Darwinian, and one sometimes hears this, because he allowed for the inheritance of acquired characteristics, not in the case of mutilations but in the cases of habit and use and disuse. But one has to keep in mind that modification is a much wider concept than adaptation, so that Darwin's claim above does not necessarily commit him to allowing processes other than natural selection a causal role in the production of adaptations.

This is indeed a contentious issue. Robert Richards (1992), for example, contends that right from the beginning Darwin allowed for a number of mechanisms to explain adaptations, but that after he hit on natural selection these mechanisms, as in the *Origin*, could no longer be central, and were retained only as "auxiliary mechanisms" (84). Richards goes on to say, however, that Darwin "asserted that acquired habit, like selection, could act more immediately to introduce adaptations; but he thought his primary device would always retain the upper hand" (n. 43). Richards goes too far, however. The passage from the *Origin* that he cites in support of his view is the following: "On the whole, I think we may conclude that habit, use, and disuse, have, in some cases, played a considerable part in the modification of the constitution, and of the structure of various organs; but that the effects of use and disuse have often been largely combined with, and sometimes overmastered by, the natural selection of innate difference" (142–143).

This passage is hardly a conclusive statement about the ability of Lamarckian mechanisms *alone* for "introducing" adaptations. In fact, Darwin in the above passage does not even mention adaptations, but only modification, which is not the same thing. In fact there is plenty of evidence that goes against Richards' interpretation. Darwin always seems to emphasize natural selection as being responsible for adaptations, possibly mixed in some instances with Lamarckian mechanisms.

For a start, there is chapter V of the *Origin*, titled "Laws of Variation." In that chapter Darwin reasserts right from the start his belief that variations are not produced by "chance" but have causes (laws), and he attempts to investigate the various causes, as well as the various relations between variations, such as what he called in chapter I "the mysterious laws of the correlation of growth" (12). It is in this chapter, on the *causes* of variations (which is a *preliminary* matter, not to be confused with adaptations *per se*), that the passage made much of by Richards is to be found. We find, for a start, that Darwin examines the effect of climate and food on the production of variations, and concludes that they probably have some effect (more in plants than in animals), but also that they "cannot have produced the many striking and complex co-adaptations of structure between one organic being and another, which we see everywhere throughout nature" (132). He then gets into the effects of use and disuse, rejecting first "that mutilations are ever inherited" (135). To use and disuse proper he admits heritable variation (134), but in each and every case of adaptation he never allows *only* use and disuse; instead he argues that natural selection had to play a big part. Darwin then turns to acclimatization, to the observation that "each species is adapted to the climate of its own home" (139). But here Darwin says "the degree of adaptation . . . is often overrated" (139). Moreover with regard to the ranges of species he takes "competition of other organic beings" to be equally or even more important than climate (140). He then questions whether habit alone could be responsible for acclimatization, calling it "mere habit," and concludes with the passage we have seen quoted by Richards. To this we should add what Darwin wrote in the summary of chapter I, the chapter titled "Variation Under Domestication," that "Variability is governed by many unknown laws, more especially by that of correlation of growth. Something may be attributed to the direct action of the conditions of life. Something must be attributed to use and disuse. The final result is rendered infinitely complex" (43). This is all about the *causes* of variation. But what has that to do with natural selection? Everything! Since "unless profitable variations do occur, natural selection can do nothing" (82).

There is further evidence that Darwin thought that natural selection must be the main explanation for adaptations. There is his "Historical Sketch" (Burkhardt *et al.* 1993, 572–576), added to the American edition of the *Origin* and to all subsequent editions. Where Darwin discusses Lamarck, he says "To this latter agency ['use and disuse or the effects of habit'] he seems to attribute all the beautiful adaptations in nature" (573). He then adds in a footnote that his grandfather "anticipated these erroneous views in his Zoonomia." Turning a little later to the author of *Vestiges*, Darwin states that his appeal to "vital forces" and "sudden leaps" "cannot . . . account in a scientific sense for the numerous and beautiful co-adaptations, which we see throughout nature" (574; cf. *Origin* 3–4).

There is more. In a letter to Gray (September 5, 1857), for example, Darwin wrote "The facts which kept me longest scientifically orthodox are those of adaptation . . . . To talk of climate or Lamarckian habit producing such adaptations to other organic beings is futile. This difficulty, I believe I have surmounted. . . . I will enclose . . . the briefest abstract of my notions on the *means* by which nature makes her species" (Burkhardt and Smith 1990, 445–446). In a letter to Hooker (July 2, 1859) Darwin wrote "to explain *adaptations* . . . this point has always seemed to me the turning point of the theory of Natural Selection" (Burkhardt and Smith 1991, 316). In a letter to William Harvey (September 20–24, 1860) he wrote "as if I said [in the *Origin*] that Natural Selection was the sole agency of modification; whereas I have over & over again, ad nauseam, directly said & by order of precedence implied (what seems to me obvious) that selection can do nothing without previous variability. . . . I consider Natural Selection as of such high importance, because it accumulates successive variations in any profitable direction; & thus adapts each new being to its complex conditions of life" (Burkhardt *et al.* 1993, 371). Finally, in a letter to Hugh Falconer (October 1, 1862) he wrote "Let me explain how it arose that I laid so much stress on Natural Selection, and I still think justly: I came to think from Geographical Distribution &c. &c. that species probably change; but for years I was stopped dead by my utter incapability of seeing how every part of each creature (a wood-pecker or swallow for instance) had become adapted to its conditions of life. This seemed to me and does still seem the problem to solve, and I think natural selection solves it, as artificial selection solves the adaptation of domestic races for man's use" (Burkhardt *et al.* 1997, 440–441).

On the other hand, interestingly, Darwin wrote in a letter to Hooker (May 29, 1860) "Why I do not believe so much as you do in Physical agencies, is that I see in almost every organism (though far more clearly in animals than in plants) *adaptation*, & this except in rare instance must, I sh^d. think, be due to selection" (Burkhardt *et al.* 1993, 230).

In sum, then, on the topic of adaptations it would seem that Darwin should be placed close to but not quite beside the modern arch-selectionist Richard Dawkins (1986, 288–318) in his argument that adaptive complexity is explicable not by habit or use and disuse or mutation or theistic design but "*only* by Darwinian selection."

At any rate, in solving the problem of adaptation, Darwin also, in his mind, solved the problem of both the origin and nature of species, and also the problem of uniting biology with the rest of natural science. Adaptation was the key in each and every case. Whether species should be so conceived, of course, is another matter. But that is a topic to be discussed in the next chapter, after we examine what Darwin did *not* include in his species concept.

Chapter 6

Not Sterility, Fertility, or Niches

We have seen in the previous chapter, specifically in the case of primroses and cowslips, that Darwin gave sterility between the two forms as one of his reasons (the last) for why they should be considered specifically distinct. But how seriously should we take this? We shall see in this chapter that we should not take it seriously at all. The other side of the coin, fertility, which I shall take to be shorthand for intrafertility, is we shall see no more a part of Darwin's species concept than sterility. In the case of primroses and cowslips, we have also seen Darwin refer to the fact that they inhabit different "stations." We shall look at what Darwin meant by this term, and we shall see that Darwin no more thought of species as inhabitants of ecological niches than he thought that they are intersterile or intrafertile. All of this is interesting for a number of reasons, one of which has to do with reconstructing Darwin's species concept, which I shall do at the end of this chapter, followed by a discussion of its strengths and weaknesses. A further reason for interest in what Darwin has to say about the above criteria has to do with modern species concepts, in other words, why Darwin would reject modern species concepts based on reproductive isolation, or fertility, or niches. Modern debaters on these topics make some interesting arguments, but Darwin has some interesting perspectives of his own which, surprisingly, seem universally to have been overlooked, perspectives that in some cases clinch the modern critiques. This chapter, therefore, will function at a number of levels.

Let us begin with sterility. We have seen in previous chapters, and we shall see again, that some modern authors believe that Darwin, in his mature writings, came close to something like the modern biological species concept. In Darwin's day, of course, the phrase "reproductive isolation mechanisms" was never used. Instead, sterility was the concept and it was a common idea. Lyell (1832) put it clear enough as "nature has forbidden the intermixture of the descendants of distinct original stocks, or has, at least, entailed sterility on their offspring, thereby preventing their being confounded together" (19). A little earlier Prichard (1813) raised the question whether nature has provided for the instinctual repugnance between species, "rather than by any more absolute decree" (10), but concluded "by the result of numerous experiments" that "the absolute sterility of such mixed offspring, must be held to be a law established in nature, and to it, rather than to any supposed agency of instinct, must be attributed the universal preservation of distinct species" (12–13). All of this finds a clear echo in Darwin's *Origin*, where he writes "The view generally entertained by naturalists is that species, when intercrossed, have been specially endowed with the quality of sterility, in order to prevent the confusion of all organic forms" (245).

As Mayr (1970) points out, sterility is only one of the reproductive isolation mechanisms in the biological species concept. According to this concept, a species is one or more actually or potentially interbreeding populations separated from other species by reproductive isolation mechanisms (12). These mechanisms, all of which "have a partially genetic base" (56), divide into two classes (57), namely, premating (seasonal and habitat isolation, ethological isolation, and mechanical isolation, an example of the latter being no sperm transfer), and postmating (gametic mortality, zygotic mortality, hybrid inviability, and hybrid sterility). The biological species concept, as Mayr further points out, is a "relational" concept, like "brother," such that it is logically impossible for the world to have only one species (13–14).

Mayr's biological species concept emphasizes what keeps species apart. There are a number of other species concepts, however, which emphasize what keeps a species together. Perhaps the most important of these is Hugh Paterson's (1985) recognition species concept. Originally Paterson characterized a species in terms of what he called a "specific-mate recognition system," defined as a signaling system between appropriate mating partners or their cells. Each species for Paterson had its own distinct specific-mate recognition system. However, due to the fact that many good species share a common specific-mate recognition system (as with several species of orchids from different genera), Paterson in his 1985 paper (which still retained the name of his older species concept) expanded the original definition to something more inclusive, namely, a distinct fertilization system (which includes the specific-mate recognition system as a subclass). (For further details on Mayr's and Paterson's species concepts, and for a detailed critique, cf. Stamos 2003, 192–205.)

A number of other biologists have defined species in terms of ecological niches, each species being kept together by a niche. I will mention a few of these when we come to look at what Darwin had to say that is relevant to the matter. What we shall see in this chapter is that not only did Darwin in his mature writings not come close to defining species in terms of sterility, or fertility, or niches, but perhaps even more importantly he raised serious difficulties for each of these approaches, difficulties that might be judged to completely undermine them, even though his concept of reproductive isolation mechanisms, fertilization systems, and ecological niches was more limited than those of the relevant modern thinkers.

A good place to begin is with what Darwin thought about dogs. It is clear that Darwin thought that the various wild species from which domestic dogs were derived were intersterile to some degree. This comes out perfectly clear in two of his letters to Lyell written in 1859 (Burkhardt and Smith 1991). In the first letter (October 25), Darwin summarizes the theory of Pyotr Pallas, which includes the view that the wild species (which Darwin expands in number) have a "tendency to sterility" that "when long domesticated they lose" (357). In the second letter (October 31), Darwin distinguishes between which parts of Pallas' theory he accepts and which parts he does not, and it is sufficient for our purposes that he writes "I must think that the sterility which they would probably have evinced if crossed before being domesticated, has been eliminated" (363).[1] In his correspondence, moreover, Darwin makes it clear that he thinks the wild parental species (plural) of domestic dogs were themselves descended from a single primitive species, in other words, that though domestic dogs as a group are polyphyletic, their parental species as a group are monophyletic. This is clear from his letter to his sister Caroline Wedgwood (November 21, 1859), in which in reply to her query on the domestic dogs issue he wrote "It is a distinct question whether these wild species have descended from one aboriginal stock as I believe has been the case" (386). Thus it would seem to be Darwin's view that intersterility was the constitutive factor that divided the aboriginal stock into a number of wild species. In other words, from the case of Darwin on dogs it certainly looks as if Darwin took reproductive isolation (in the strict sense of sterility) to be a constitutive factor in the ontology of species (combined of course with other factors such as common descent). This conclusion appears only stronger if we take what Darwin says about primroses and cowslips examined at the beginning of the previous chapter.

And indeed those few authors who have interpreted Darwin in his mature period as a species realist of some sort have advocated this conclusion (regarding sexual taxa) in one way or another. Among the strategy theorists, it is implied by Beatty (1985) in his interpretation of Darwin as believing that what his fellow naturalists called "species" were really "chunks with[in] the genealogical nexus of life" (278). In my first paper on Darwin (Stamos 1996), I explicitly held that Darwin's mature species concept involved sterility/fertility to a high degree, such that "It is the evidence that these two criteria ['interspecific sterility and intraspecific fertility'] are *almost universal* that allows Darwin to steer a course right down the middle between species nominalism and species essentialism; in other words, on the one hand it allows for species (for the most part) to be objectively real, while on the other hand it allows for the indefinite mutability of species—both of which are required for his theory" (140).

But it is the earlier theorists, namely, Ghiselin and Kottler, who provide the most interesting arguments and examples in this vein. What we shall see in this chapter, however, is that if we take a closer and more critical look at their examples than they did, then we shall gain insights that in turn will require us to reinterpret what Darwin says about dogs above and primroses and cowslips in the previous chapter.

Beginning with Ghiselin (1969), he focuses on a paper of Darwin's on primroses and cowslips (Darwin 1869) surprisingly not included in Barrett's (1977) anthology. According to Ghiselin, in this paper Darwin

> demonstrates that the intermediate form, or oxlip, is a sterile hybrid, and supports this inference by showing that the oxlip occurs where the parent species are present, but not otherwise. The third species is shown to be sterile when crossed with the others, and to be distinct in morphology and in geographical range. In summing up his observations, Darwin says that the various forms "are all descended, from the same primordial form, yet, from the facts which have been given, we may conclude that they are as fixed in character as the very many other forms which are universally ranked as species. Consequently they have as good a right to receive distinct specific names as have, for instance, the ass, quagga, and zebra. [100]

"When we realize that the analogy with horses," continues Ghiselin, "is an oblique reference to mules, we see that Darwin was stressing reproductive distinctness" (100). Ghiselin nevertheless qualifies this interpretation on the very next page: "Let it not be thought, however, that Darwin supported the biological species definition in its strictly modern sense. There is no solid evidence that he conceived of species as reproductively isolated populations. His emphasis lay more with the distinctness of the individuals in different species in terms of their biologically important characteristics, and also with the genealogical interrelationships of the individuals within each species" (101).

It must be somewhat odd, however, if Ghiselin's interpretation is correct, that Darwin (as we shall see) would sometimes stress reproductive distinctness and sometimes not, and with no apparent rhyme or reason. Certainly a more charitable interpretation is desirable. And in fact it is possible to reinterpret Ghiselin's evidence as not stressing reproductive distinctness at all, which would make Darwin more internally consistent. This is the interpretation of Darwin's mature species concept that I wish to develop and defend here, namely, that for Darwin sterility, and its correlative fertility, are not constitutive of species distinctness and species unity respectively, but instead merely serve as probable evidence of such. (I shall return later to what Ghiselin above calls Darwin's "oblique reference to mules.")

The evidence supporting this radical reinterpretation is pervasive. There is, first of all, Darwin's repeated efforts in the *Origin* toward the conclusion that sterility between species "is not a special endowment, but is incidental on slowly acquired modifications, more especially in the reproductive systems of the forms which are crossed" (272). As Darwin more fully put it,

> The importance of the fact that hybrids are very generally sterile, has, I think, been much underrated by some late writers. On the theory of natural selection the case is especially important, inasmuch as the sterility of hybrids could not possibly be of any advantage to them, and therefore could not have been acquired by the continued preservation of successive profitable degrees of sterility. I hope, however, to be able to show that sterility is not a specially acquired or endowed quality, but is incidental on other acquired differences. [245]

Given, then, that in Darwin's view sterility between species cannot be a direct consequence of natural selection, unlike other characters such as beaks and eyes, so that it is therefore not an adaptation, and given what we have seen in the previous chapter, it is legitimate to hypothesize that in Darwin's view any property that is not an adaptive consequence of natural selection is not a species-specific property, it is not part of what a species is.[2]

This conclusion seems to further recommend itself by the fact that Darwin allows for intraspecific sterility, not just in the sense of sterile castes but more importantly between fertile individuals of the same species. In the *Origin* Darwin gives a number of examples of conspecific varieties with "a certain amount of sterility" (269) between them. Perhaps the most interesting case is the white and yellow varieties of nine species of the genus *Verbascum* as studied "by so good an observer and so hostile a witness, as Gärtner," hostile because like so many others he considered "fertility and sterility as safe criterions of specific distinction" (270). With these varieties no one (including Darwin) doubted that they are varieties, since the yellow and white varieties of each species "present no other difference besides the mere colour of the flower; and one variety can sometimes be raised from the seed of the other" (271). What is so remarkable about them is that within each species there was found to be much less fertility when the white and yellow varieties were crossed compared to when the crosses were confined to within each of the two varieties (cf. Stauffer 1975, 406–407; Darwin 1876c, 258–259). Similar results are reported by Darwin from the experiments by Gärtner on maize, Girou de Buzareingues on gourds, Kölreuter on tobacco, and Darwin's own observations on hollyhocks, which he suspects "present analogous facts" (271).[3]

Of further interest are two papers of Darwin's, one on the sexual relations within and between the dimorphic forms (short-styled and long-styled) of primroses and cowslips (Darwin 1862b), the other on the sexual relations within and between the trimorphic forms (short-styled, mid-styled, and long-styled) of the purple loose-strife, *Lythrum salicaria* (Darwin 1865). In the former case, Darwin wrote in conclusion that "The simple fact of two individuals of the same undoubted species, when homomorphically united [long-styled with long-styled, short-styled with short-styled], being as sterile as are many distinct species when crossed, will surprise those who look at sterility as a special endowment to keep created species distinct" (Darwin 1962b, 61). In the latter case, the case of *Lythrum salicaria,* "in each of the three forms the fertile unions are all *heteromorphic,* the appropriate pollen coming from the stamens of corresponding length borne by the other two forms, and . . . the *homomorphic* unions of the females with their own two sets of males are always more or less sterile" (Darwin 1865, 120). In the concluding paragraph of the paper he wrote that "differences in sexual nature have been thought to be the very touchstone of specific distinction. We now see that such sexual differences—the greater or less power of fertilizing and being fertilized—may characterize and keep separate the coexisting individuals of the same species, in the same manner as they characterize and have kept separate those groups of individuals, produced from common parents during the lapse of ages or in different regions, which we rank and denominate as distinct species" (128). And in the concluding paragraph of a much later paper (Darwin 1880), he wrote that "mutual sterility is no safe and immutable criterion of specific difference. We have, however, much better evidence on this head, in the fact of two individuals of the same form of heterostyled plants, which belong to the same species as certainly as do two individuals of any species, yielding when crossed fewer seeds than the normal number, and the plants raised from such seeds being, in the case of *Lythrum salicaria*, as sterile as are the most sterile hybrids" (220).[4]

Moreover we have already seen in the case of domestic dogs and cattle, and also in the theoretical case of convergence at the species level, that Darwin did not take interbreeding or interfertility as a criterion of species unity. And given this fact and the previous, it becomes difficult to maintain the view that Darwin's mature species concept included either reproductive isolation or interbreeding as constitutive of species.

If not constitutive, Darwin nevertheless held, as he put it in the *Origin*, both that the fertility between conspecific varieties is "almost invariably the case" (268) and that sterility barriers between different species "cannot, under our present state of knowledge, be considered as absolutely universal" (254). That neither are absolutely universal but almost universal is quite sufficient for them to serve as good preliminary evidence of species distinctness, even though, as we have seen, they are not constitutive properties of species.

The same conclusion may be inferred from Darwin's discussion in the *Origin* on "systematic affinity," which he defines as "the resemblance between species in structure and in constitution, more especially in the structure of parts which are of high physiological importance and which differ little in the allied species" (256). Of the covariance between the degree of systematic affinity and the degree of hybridity, Darwin affirms that they do largely covary, but he then immediately adds that "the correspondence between systematic affinity and the facility of crossing is by no means strict" (257), indeed that "the capacity in any two species to cross is often completely independent of their systematic affinity" (258). Sometimes, as Darwin points out, there is a very high degree of systematic affinity between two species but very little if any interfertility, while sometimes there is a very high degree of interfertility but low systematic affinity. As such, then, systematic affinity can do no more than merely serve as preliminary evidence of interfertility. Thus, to apply once again the modern distinction employed in Chapter 4, intraspecific fertility and interspecific sterility are for Darwin fallible diagnostic criteria of species distinctness, and they are fallible because although usually found in and between distinct species they are not constitutive characters of species.

Instead of fertility and sterility, then, the characters that do count for Darwin as being constitutive of species are relatively constant and distinct characters (morphological, ecological, and behavioral), characters moreover of high physiological importance fixed as such by natural selection, all of which we have seen in chapter 5. Sterility between species, on the other hand, is not a character produced by natural selection, it is not adaptive, and therefore should not count in a *vera causa* classification. Instead for Darwin sterility should only serve as preliminary evidence.

Much the same goes for fertility, although this is far more problematic and we have to be very careful here. In the *Origin* Darwin refers to hybrid vigor, inbreeding depression, and the necessity of a certain amount of crossing for hermaphrodites (267), and it might naturally be supposed that he thought of this as adaptive, as something selected for. However, only a few pages later, Darwin seems to assert the opposite and to view fertility in the same category as sterility, not as something selected for but as a consequence of selection:

> I do not think that the very general fertility of varieties can be proved to be of universal occurrence, or to form a fundamental distinction between varieties and species. The general fertility of varieties does not seem to me sufficient to overthrow the view which I have taken with respect to the very general, but not invariable, sterility of first crosses and of hybrids, namely, that it is not a special endowment, but is incidental on slowly acquired modifications, more especially in the reproductive systems of the forms which are crossed. [272]

At any rate, to understand Darwin on this matter we have to make a distinction between fertilization mechanisms, which as we have seen in the previous chapter on the topic of orchids Darwin clearly thought were adaptations, and fertility *per se*, whether self-fertility in hermaphrodites or intraspecific fertility. Just because Darwin thought fertilization mechanisms were adaptations, we cannot therefore assume that he also thought that fertility *per se* was adaptive. On the contrary, there is plenty of evidence that should direct us toward the opposite conclusion, that fertility *per se* was not for Darwin adaptive, or often not adaptive. Some of that evidence comes from what Darwin would later claim about self-sterility in plants. Darwin was well aware that this was a matter of degree throughout the plant kingdom, the opposite end being the so-called Darwin-Knight law (that no plant self-fertilizes perpetually). In an interesting discussion, Darwin (1876b) points out that it would be natural to suppose that self-sterility had been gradually evolved by natural selection. However, he then goes on to reject this view, his main reason being "the immediate and powerful effect of changed conditions in either causing or in removing self-sterility. We are not therefore justified in admitting that the peculiar state of the reproductive system has been gradually acquired through Natural Selection; but we must look at it as an incidental result dependent on the conditions to which the plants have been subjected, like the ordinary sterility caused in the case of animals by confinement, and in the case of plants by too much manure, heat, etc." (346). By parity of reasoning, Darwin may well have held the same view about much of intraspecific fertility, that it is an incidental result dependent on the conditions of life. We have to recall that in Darwin's view natural selection produces adaptations only for the good of the individual, a claim repeated throughout the *Origin* (e.g., 83–86), or of the community only when that contributes to the good of each individual in the community (87), where by "good" Darwin clearly meant "welfare" (127). Darwin, therefore, might well have held the view that, just as interspecific sterility does the individual no good, so likewise intraspecific fertility does or often does the individual no good, so that combined with the evidence from the effect on reproductive systems by the conditions of life, there is no good reason to suppose that fertility is always selected for.

What further supports this interpretation is Darwin's claim that fertility can actually oppose natural selection. In the *Origin* he says "Intercrossing plays a very important part in nature in keeping the individuals of the same species, or of the same variety, true and uniform in character" (103), to which he adds "The process [of natural selection] will often be greatly retarded by free intercrossing. Many will exclaim that these several causes are amply sufficient wholly to stop the action of natural selection. I do not believe so" (108).

We should also return to Darwin's (1859) prime example of the white and yellow varieties of the nine species of the genus *Verbascum*. It was further found by Gärtner that "when yellow and white varieties of one species are crossed with yellow and white varieties of a *distinct* species [of *Verbascum*], more seed is produced by the crosses between the same coloured flowers, than between those which are differently coloured" (270–271; cf. Stauffer 1975, 406–407; Darwin 1876c, 258–259).

With this new interpretation, that for Darwin sterility and fertility are generally good evidence of species distinctness but are not constitutive of species distinctness, we may look with new eyes at the evidence adduced in favor of the interpretation that Darwin's mature species concept involved to some extent both "reproductive distinctness," as Ghiselin put it above, and "the genealogical interrelationships of the individuals within each species." With Ghiselin's example of Darwin's (1869) reference to the ass, quagga, and zebra, it is difficult, especially following the analysis above, to see how it is an oblique reference to mules, and therefore as stressing reproductive distinctness. Whatever the degree of sterility or fertility between these three species, Darwin need only have been referring to their relatively constant and distinct characters (as indeed he does in Ghiselin's quotation). Interestingly, in one of his later papers Darwin (1880) remarked on the perfect fertility between the common and Chinese goose, which, because of their distinctness in many characters, elucidated by Darwin in his paper, some authors had placed in different genera. But because of the discovered perfect fertility between them, one author in particular had suggested that the Chinese goose is only a domestic variety of the common goose. "But it would, I believe," replied Darwin, "be quite impossible to find so many *concurrent and constant* points of difference as the above, between any two domesticated varieties of the same species. If these two species are classed as varieties, so might the horse and ass, or the hare and rabbit" (219–220, italics in original). Far from being an oblique reference to reproductive distinctness (especially since in the goose example there is nothing comparable to mules), the reference here is to constant character differences, precisely in accordance with the analysis above.[5]

Perhaps more challenging than the evidence brought forward by Ghiselin is the evidence brought forward by Kottler (1978), according to whom "passages from the *Origin* and later works, as well as Darwin's taxonomic practice, suggest that he did appreciate the importance of biological criteria, including reproductive isolation" (292). Kottler's main piece of evidence is Darwin's reference to sibling (cryptic) species (both terms are of twentieth century vintage), species that are so morphologically similar that their minor differences were either missed or ignored until it was discovered that they were reproductively isolated from one another (hence they were originally thought to be one species). In particular, Kottler focuses on Darwin's repeated reference to willow wrens, referred to especially in his unfinished *Natural Selection* and also in his review of H.W. Bates' paper on mimicry in butterflies.

My difficulty with Kottler's use of these references, however, is that there is no mention by Darwin of any kind of reproductive isolation, only the maintenance of distinct characters. In the passage from *Natural Selec-*

tion that Kotter quotes (1978, 280 n. 9), Darwin wrote "In species we should remember how extremely close some undoubtedly distinct forms are, as many plants, & as in some of the willow wrens, which are so close that the most experienced ornithologists can hardly distinguish them except by their voice, & the materials with which they line their nests; yet as these wrens inhabit the same country & always exhibit the same difference, no one can doubt that they are good species" (Stauffer 1975, 99).

Of course one might *infer* that Darwin is implicitly referring to reproductive isolation here. One area of support comes from his transmutation notebooks, specifically Notebook B (Barrett *et al.* 1987, 224), wherein Darwin gave one of his definitions of species, which implicitly includes willow wrens: "Definition of Species: one that remains at large with constant characters, together with beings of very near structure.—Hence species may be good ones & differ scarcely in any external character:—For instance two wrens forced to haunt two islands one with one kind of herbage & one with other, might change organization of stomach & hence remain distinct" (213). Given that on the previous page Darwin wrote that "between Species from moderately distant countries, there is no test but generation . . . whether good species" (212), it is with good reason that Darwin scholars such as Kottler (1978, 279–280), Sulloway (1979, 29), and Mayr (1982, 266) use this passage to interpret Darwin's early species concept as including sibling species.

In further support of Kottler's interpretation of the willow wren reference in *Natural Selection*, it should be pointed out that the concept of sibling species is also to be found in the *Origin*, where Darwin affirms that "many species there are, which, though resembling each other most closely, are utterly sterile when intercrossed" (268). Moreover Kottler's interpretation may seem natural given the modern prevalence of the biological species concept, with its inclusion of sibling species. But of course one must always be careful not to allow the classic mistake of letting a highly influential modern scientific theory or definition influence one's historical interpretation of an earlier scientist. And this is possibly what we have here.[6]

Although to modern eyes it might look like Darwin, in his reference to willow wrens in *Natural Selection*, is referring to sibling species and thereby implicitly employing them in his mature species concept, there is, again, no actual reference to reproductive isolation. And given the extended analysis in the present chapter on Darwin's rejection of sterility as constitutive of specific difference, the only warranted conclusion is that the implication of reproductive isolation is just not there. All that mattered to Darwin in terms of his mature species concept, as follows from the present chapter and my analysis in the previous chapter, was that the two species have at least one character difference, in particular, a different adaptation, and that the difference has persisted in spite of the two species overlapping in range. Indeed, in *Natural Selection* this comes out at the end of the chapter quoted from above. Therein Darwin distinguishes between species and varieties, the former in his view "being much more constant in all their characters," this constancy being due mainly to "the several causes of variability having acted less energetically on the two species under comparison than on the one species yielding the two or more varieties; and partly to the characters of the two species having been long inherited, & by this very cause having become more fixed" (Stauffer 1975, 165). Later in the same work, in his chapter on natural selection, Darwin connects this constancy of characters with natural selection:

> In regard to the difference between varieties and species, I may add that varieties differ from each other & their parents, chiefly in what naturalists call unimportant respects, as size, colour, proportions &c; but species differ from each other in these same respects, only generally in a greater degree, & in addition in what naturalists consider more important respects. But we have seen in Ch. IV, that varieties do occasionally, though rarely, very [*sic*, vary] slightly in such important respects; and in so far as differences in important physiological characters generally stand in direct relation to different habits of life, modifications however slight in such characters would be very apt to be picked out by natural selection & so augmented, thus to fit the modified descendants from the same parent to fill as many & as widely different places in nature as possible. [Stauffer 1975, 244][7]

Indeed Darwin is not even committed to reproductive isolation keeping two closely related, overlapping forms distinct. Although in *Natural Selection* he admits that intercrossing "will obviously retard, perhaps obliterate, the process of selection by dragging back the offspring of a selected variety towards its parental type" (Stauffer 1975, 255), he nevertheless immediately adds that this is a topic filled with "doubt." As he put it in the concluding section of his chapter on natural selection, "Intercrossing will prevent or retard the process of natural selection; but here we are involved in much doubt." The doubt surrounds the weighing of several contingencies responsible for "structural change" (271), or, as he a little later put it, "the several contingencies favorable to natural selection." "I am inclined," wrote Darwin, "to rank changed relations or associations between the inhabitants of a country from opening up new places in its polity, as the most important element of success." Equally significant is that a few lines later he wrote that "A diminished amount of intercrossing is probably the least important element" (273).[8]

The second piece of evidence adduced by Kottler (1978) is from Darwin's review of Henry Walter Bates' (1862) paper on mimicry in butterflies. In the second last paragraph of his otherwise favorable review, Darwin (1863c) remarks that Bates "ought . . . to have given in every case his reasons in full for believing that the closely allied and co-existing forms, with which his varieties do *not* pair, are not distinct species. Naturalists should always bear in mind such cases as those of our own willow wrens, two of which are so closely alike that experienced ornithologists can with difficulty distinguish them, excepting by the materials of which their nests are built; yet these are certainly as distinct species as any in the world" (92). Kottler's footnote in which this passage is quoted and discussed (Kottler 1978, 280 n. 9) is a continuation of his discussion on Darwin's "biological" species concept in the transmutation notebooks and Darwin's example of willow wrens, which according to Kottler illustrates Darwin's inclusion of sibling species in that concept. Although Kottler in his footnote seems more concerned to point out that Darwin in the final page of his review was prodding Bates to provide more evidence of "preferential mating," mating within though not between overlapping varieties of the same species, it is obvious that Kottler's use of the willow wrens reference is meant to support his claim that even in Darwin's mature period his realist species concept included reproductive isolation to some degree.

When I read the above passage from Darwin's review, however, I am instead reminded of a particular passage from the *Origin* that we have seen before, *viz.*, "Those forms which possess in some considerable degree the character of species, but which are so closely similar to some other forms . . . that naturalists do not like to rank them as distinct species, are in several respects the most important for us. We have every reason to believe that many of these . . . closely-allied forms have permanently retained their characters in their own country for a long time; for as long, as far as we know, as have good and true species" (47). This passage, it seems to me, has those like Bates written all over it. Although Bates adduced evidence of preferential mating in overlapping conspecific varieties, he nevertheless categorized those varieties as varieties, not as species. This is especially clear in the case of *Mechanitis polymnia*, which Bates says helped destroy his belief in the constancy of species (530), and which Darwin accordingly refers to in the third last paragraph in his review as a case that "well deserves study: after describing its several varieties, Mr. Bates adds, 'these facts seem to teach us that, in this and similar cases, a new species originates in a local variety . . . from one variable and widely distributed species'" (92). Bates tells us that *M. polymnia* is highly polymorphic (500 n*) and generally presents only one local form in a district (531 n*). What is especially interesting about *M. polymnia*, however, is not that some of its varieties, for example, the predominant *M. Egaënsis* (Bates normally used binomials for varieties), exhibit preferential mating (529), but rather its varieties (or rather the members of those varieties) are uniquely adaptive. The unique adaptation(s) of each variety, however, is not that of mimicry, which Bates, of course, thought of as an adaptive product of natural selection (508). *M. polymnia* is one of those species that is mimicked (502, 503, 564, 566). Nevertheless Bates thinks of its varieties as differently adapted, by selection, "the selected ones being different in different districts," though "What these [the adaptive] conditions were, or have been, was not revealed by the facts" (511).

What Darwin is doing at the end of his review, then, it seems to me, is challenging Bates' species concept on a matter of theoretical consistency. At one point Bates himself seems to equate a new species with a new adaptation, when he says the case of mimetic species "offers a most beautiful proof of the truth of the theory of natural selection. It also shows that a new adaptation, or the formation of a new species, is not effected by great and sudden change, but by numerous small steps of natural variation and selection" (513). But in practice, as in the case of *M. Egaënsis*, Bates still considered it a variety of *M. polymnia*, rather than a species in its own right, in spite of its having a unique adaptation. Part of his reason was the existence of connecting forms (530), which unlike Darwin in cases such as primroses and cowslips Bates repeatedly took as a criterion of conspecific inclusion (501, 532, 546, 555). The other reason was that it did not meet what Bates considered to be the test of species attainment: "The new species cannot be proved to be established as such, until it be found in company with a sister form which has had a similar origin, and maintaining itself perfectly distinct from it" (530). What matters to Darwin, on the other hand, is the permanence-of-characters criterion, characters moreover selected and made relatively constant by natural selection, and this is evident in the passage quoted above from Darwin's review. It is evident in his reference to willow wrens, and also in his own examples of preferential mating, such as "that two herds of differently coloured deer long preserved themselves distinct in the New Forest" (92).[9] The preferential mating, however, is not what Darwin's species concept is about, nor his challenge to Bates, since preferential mating would for Darwin be no more adaptive by selection than would sterility. Instead his challenge to Bates is to explain why his varieties, each with a unique adaptation, are not classified by Bates as distinct species. Of course other interpretations of Darwin's challenge to Bates are possible, but the above interpretation is the most consistent with everything else we have seen and is therefore the one to be preferred.

Having rejected sterility and fertility as parts of Darwin's species concept, it remains to consider whether he would have accepted a niche component. A number of modern biologists have defined species either in terms of the occupants of a niche or as having a niche component. Ehrlich and Raven (1969), for example, impressed by the weakness of gene flow within many good species, by the amount of gene flow between many good species, and by the sheer lack of it in asexual species, defined species in terms of "similar selective regimes" (61). Van Valen (1976) is another example motivated by much the same reasons. For him a species is a lineage or set of lineages that occupies an "adaptive zone" minimally different from any other lineage in its range (70). Others still, such as Simpson (1961), defined a species not as occupying a niche throughout its entire existence (recall that Simpson conceived of species as vertical entities), so that the demarcation of a species vertically would not be by a niche, but rather only as having at any one time a unique adaptive "role," which he equated with a "niche" (154–155). With each of these biologists the niche is conceived to be a property of the environment. But in some species concepts the niche concept is more modern, conceiving of a niche as the actual or potential resource utilization of a population or species, so that the niche is not a property of the environment (hence no empty niches) but rather a property of the population or species. Thus Templeton (1989), for example, defines a species as an evolutionary lineage "having the potential for phenotypic cohesion through intrinsic cohesion mechanisms" (12), and he divides these cohesion mechanisms into basically two categories, the one being the exchange of genes via sexual reproduction, the other being the fundamental (potential resource utilization) niche of the species. (For further details on ecological niche species concepts as well as on ecological niche concepts, cf. Stamos 2003, 144–156.)

Did Darwin have any of this in his species concept, and would he have endorsed any of this? Although the actual use of the word "niche" as a concept in ecology did not begin until the early 1900s, the *concept* was clearly in circulation in Darwin's time and even a little before it. Lyell (1832), for example, used the phrase "the economy of Nature" (42), and after quoting from an 1820 essay by Augustin Pyramus de Candolle, in which the latter distinguished "stations" from "habitations," Lyell wrote

> we may remind the geological reader that station indicates the peculiar nature of the locality where each species is accustomed to grow, and has reference to climate, soil, humidity, light, elevation above the sea level, and other analogous circumstances; whereas by habitation is meant a general indication of the country where a plant grows wild. Thus the *station* of a plant may be a salt-marsh, in a temperate climate, a hill-side, the bed of the sea, or a stagnant pool. Its *habitation* may be in Europe, North America, or New Holland between the tropics. . . . The terms thus defined, express each a distinct class of ideas, which have been often confounded together, and which are equally applicable in zoology. [69]

Not much later Darwin (1859) would write, as we have seen, of the "somewhat different stations" of primroses and cowslips (49), and more generally of "each place in the economy of nature," some of which are "unoccupied" and which natural selection tries to "fill up" (102; cf. 104). In turn Bates (1862) would write of "The two sets of forms [that] seem to agree . . . in habits, and apparently occupy the same sphere in the economy of nature in their respective countries" (498).

But niches played no role in the ontology of species for Darwin, and his reasons were quite simple. As Darwin put it in the *Origin*,

> We can clearly understand why a species when once lost should never reappear, even if the very same conditions of life, organic and inorganic, should recur. For though the offspring of one species might be adapted (and no doubt this has occurred in innumerable instances) to fill the exact place of another species in the economy of nature, and thus supplant it; yet the two forms—the old and the new—would not be identically the same; for both would almost certainly inherit different characters from their distinct progenitors. [315]

We can see here that Darwin thought of species as *forms*, though of course as plastic, rather than simply as the occupants of a niche, but also that he thought of them in terms of adaptations, so that even if a very different lineage would come to occupy a niche vacated for whatever reasons by a former lineage, the two populations (horizontally conceived) would not have the same suite of adaptations because of their different evolutionary histories.[10]

A further reason given by Darwin is that defining species in terms of niches, rather than adaptations, makes a mess of biogeography and the related issues of migration and evolutionary history, that is, phylogeny in the most inclusive sense. In the *Origin* Darwin makes this point with regard to the relationship between the endemic Galapagos species and the related South American species. As he put it,

> why should the species which are supposed to have been created in the Galapagos Archipelago, and nowhere else, bear so plain a stamp of affinity to those created in America? There is nothing in the conditions of life, in the geological nature of the islands, in their height or climate, or in the proportions in which the several classes are associated together, which resembles closely the conditions of the South American coast: in fact there is a considerable dissimilarity in all these respects. On the other hand, there is a considerable degree of resemblance in the volcanic nature of the soil, in climate, height, and size of the islands, between the Galapagos and Cape Verde Archipelagos: but what an entire and absolute difference in their inhabitants! . . . I believe this grand fact can receive no sort of explanation on the ordinary view of independent creation; whereas on the view here maintained, it is obvious that the Galapagos Islands would be likely to receive colonists, whether by occasional means of transport or by formerly continuous land, from America; and the Cape de Verde Islands from Africa; and that such colonists would be liable to modification;—the principle of inheritance betraying their original birthplace. [398–399]

In rejecting niches as part of what a species is (although he probably didn't realize he was doing that), Darwin was indeed avoiding a dead end, for not only are there at least four main niche concepts in circulation in modern biology, but whichever concept one chooses the concept of niche is a notoriously problematic concept, such that some have despaired of the concept having any objectivity at all. For example, in cases of pronounced sexual dimorphism, such as male and female woodpeckers where the males have larger beaks and feed in different parts of the trees than females, or in cases of ontogenetic dimorphism, such as caterpillar/butterflies where different ontogenetic stages occupy very different niches, if in each case one unites the niches into a single niche, it is obvious that one is presupposing the species and then inferring the niche, whereas if one began with the niches and defined species in terms of niches, then one would have to divide the males and females of many species into different species, as well as different ontogenetic stages. Moreover, niche talk is a highly subjective exercise. One can specify niches *ad infinitum*, depending on the imagination of the theorist and the degree of vagueness or precision desired. But that does not make niches objective. Similar problems attend the distinction between actual and potential niches. The potential niche of a population or species is just as conjectural and vague and indeterminable as the concept of environmental niche, whereas if one confines oneself only to the actual niche, for example, in the sense of resource utilization, then what are clearly different species might well utilize the same resources. In short, resolving the species problem into the niche problem does not solve the species problem, since the niche problem has too many empirical and conceptual problems to ever be resolved, such that many ecologists themselves have despaired of the concept altogether (cf. Stamos 2003, 153–165). Darwin was indeed prescient to focus on forms rather than on niches when it came to the ontology of species.

In rejecting niches, we can figure out what Darwin would say to modern species concepts that employ niches. Likewise, we can figure out what Darwin would say to modern species concepts that employ sterility or fertility.

In the case of sterility, we know from what Darwin wrote in reply to Huxley. In many of his post-*Origin* writings on Darwin's theory of evolution, Huxley (e.g., 1860b, 555) distinguished between what he called "morphological species" (species based on morphology) and "physiological species" (species based on interbreeding) and then went on to claim that, in accordance with the inductive standards of science, selection has proven itself "over and over again" capable of producing morphological species, but as for physiological species "there is no positive evidence at present that any group of animals has, by variation and selective breeding, given rise to another group which was even in the least degree infertile with the first" (567–568). To be fair, though, Huxley added his belief that experiments in a few years will "very probably obtain the desired production of mutually more or less infertile breeds from a common stock" (568), as well as his belief that "nature does make jumps now and then" (569), summing the situation up with the claim that Darwin's theory "is as superior to any preceding or contemporary hypothesis . . . as was the hypothesis of Copernicus to the speculations of Ptolemy" (569).

The repeated claim by Huxley that Darwin's theory fell short of empirical evidence on the production of species was a constant thorn in Darwin's side, and for a brief while Darwin acquiesced to a certain degree. For example, in a letter to Huxley (January 14, 1862) Darwin countered with "you never allude to the excellent evidence of *varieties* of Verbascum & Nicotiana being partially sterile together," to which he added a few lines later "Do oblige me by reading latter half of my Primula paper in Lin. Journal for it leads me to suspect that sterility will hereafter have to be largely viewed as an acquired or *selected* character" (Burkhardt *et al.* 1997, 19). Nevertheless, as we have seen earlier in this chapter (note 3), Darwin shortly after gave up the idea that selection could produce sterility and permanently reverted to his former view in the *Origin*. Consequently for Darwin, Huxley's physiological species were a red herring (in fact were not necessarily species at all), and from 1863 onward Darwin's claim against Huxley's view was not only that sterility is an incidental effect of selection rather than a product of selection, but that sterility could not possibly be a species character since in some clearly good species, for example, in *Verbascum* and in his later example of *Lythrum salicaria* (Darwin 1865), there is sterility *within* the species. Consequently in early 1863 in a letter to Huxley (January 10) we can see Darwin's defiance of Huxley, where he indeed uses varieties of *Verbascum* along with the kind of arguments we would later see in *Variation*, promising that he will "not hold my tongue" (Burkhardt *et al.* 1999, 29). Almost four years would go by with apparently no debate between Darwin and Huxley on the topic, but when it came up again, Huxley apparently stuck to his old line, to which Darwin in a letter to Huxley (January 7, 1867) replied "Nature never made species mutually sterile by selection, nor will men" (Kottler 1985, 407).

Darwin's arguments against sterility as a species criterion prove to be highly significant when we consider the modern biological species concept. Critics of this concept have routinely argued that it does not work well in plants, not only that it does not at all apply to asexual organisms but that many conspecific sexual populations have little or no gene flow between them so that it cannot be gene flow that gives them their cohesion (e.g., Ehrlich and Raven 1969; Levin 1979), that between many taxonomically good species there is significant gene flow (e.g., Van Valen 1976), and that asexual organisms generally divide into good species as well as if not better than sexual species (e.g., Templeton 1989). Especially interesting is Hugh Paterson's (1985) argument that the isolating mechanisms of the biological species concept, in particular the premating mechanisms, could not possibly have been evolved by natural selection (a view denied by Dobzhansky but conceded by Mayr), so that they should be considered merely as effects rather than as mechanisms proper (22). While Paterson's argument is closest to Darwin's, what needs to be added to all these criticisms to arguably clinch the case is Darwin's focus on sterility *inside* a species, whether *between* what are unequivocally varieties of the same species, as with *Verbascum*, or *within* what are unequivocally varieties of the same species, as with each of the three forms of *Lythrum salicaria*. Since sterility is a major kind of reproductive isolation, reproductive isolation could not possibly be, for Darwin, the defining characteristic of species.

Does this mean that fertility must go out the window too? Fertility is not simply the other side of the coin, since it is not affected by all the problems presented by the sterility criterion, such as Darwin's *Lythrum* example, in which the three forms of *Lythrum salicaria* comprise together a fertile whole. Interestingly, Paterson (1985) thinks that his recognition species concept, which defines a species as having a unique fertilization system, is closest to Darwin's, since, unlike reproductive isolation mechanisms (so-called), fertilization mechanisms are indeed adaptations produced by natural selection (25). Indeed Paterson goes on to claim that, in the first sentence of his orchid book, Darwin (1862a) "expresses a view in perfect agreement with the recognition concept: 'The object of the following work is to show that the contrivances by which the orchids are fertilized are as varied and almost as perfect as any of the most beautiful adaptations in the animal kingdom; and, secondly, to show that these contrivances have for their main object the fertilization of the flowers with pollen brought by insects from a distant plant'" (25).

Paterson's species concept has gained a significant following (cf. Lambert and Spencer 1995), but can he truly count Darwin as a precursor? I have to say no, for the simple reason that although fertilization mechanisms are indeed adaptations and as we have seen adaptations are the key for Darwin, fertilization mechanisms are not the only adaptations for Darwin (or for anyone else for that matter), so that it is possible for him that two populations could have the exact same fertilization mechanisms and yet be distinct species. And indeed we see good evidence for this in Darwin's writings. In the *Origin* Darwin discusses the example of two species of *Crinum* experimented on by Herbert, in which he found, in Darwin's words, that "some hybrids are perfectly fertile" (250). Darwin also discusses the examples of "certain species of Lobelia" and "all the species of the genus Hippeastrum," which, he says, "can be far more easily fertilised by the pollen of another and distinct species, than by their own pollen" (250). He also gives the example of *Verbascum* (270–271), as we have seen earlier in this chapter. Moreover, in his long chapter on hybridism in *Natural Selection* he gives numerous examples among plants and animals and with varying degrees of fertility (Stauffer 1975, 388–462).

So neither sterility, fertility, nor niches defined species according to Darwin. Having established that conclusion, and that the key for Darwin is adaptations, we can finally provide a definition of Darwin's implicit species concept, which as we have seen shows a remarkable consistency throughout his mature writings. Accordingly I would say that in Darwin's view *a species is a primarily horizontal similarity complex of organism morphologies and instincts distinguished at any one horizontal level primarily by relatively constant and distinct characters of adaptive importance from the viewpoint of natural selection, and secondarily (though not always applicable) by common descent and intermediate gradations.*

It is interesting to notice some of the implications that follow from this species concept for speciation. Speciation is a concept, of course, which is dependent on a concept of species. Without a concept of species, speciation is a vacant concept. It follows, then, that what is speciation according to one species concept might not be speciation according to another species concept. It all depends on the particular features of each species concept. At the end of chapter 3 we have seen some of the implications for speciation that follow from taking species to be primarily horizontal entities. In the present chapter we have seen that sterility and fertility are not for Darwin species characters. This has an important implication for anagenesis, speciation in a single evolutionary line. In phylogenetic systematics, in particular cladism, the possibility of anagenesis is rejected as a matter of principle. Instead only cladogenesis, branching speciation, is allowed. One reason for this is a distaste for the concept of similarity (e.g., Ridley 1989). The rejection of similarity, however, as we have seen in chapter 4, creates serious problems for phylogenetic systematics, absurdities that we have further seen at the end of chapter 3, such that some taxonomists who generally adhere to the principles of cladism have found it necessary to let similarity back in, through the back door as I like to put it, rejecting what has been called "Hennigian extinction," for example.

We can see in Darwin's species concept, however, in particular on the issue of extinction, that similarity is not rejected, nor is it sheepishly or reluctantly accepted simply to avoid absurdities. Instead it is robustly affirmed, not in the facile and arbitrary sense of overall similarity, but in the sense of being objectively *bounded*, both by the horizontal dimension and by common descent, but most importantly by adaptive characters produced by natural selection. Darwin, for that reason alone, would not reject anagenesis.

A further reason for why anagenesis is rejected in modern phylogenetic systematics, it seems to me, is due to the vestige of the sterility criterion in the work of the founder of cladism, Willi Hennig. Hennig (1966, 51–52, 58) recognized the reality of species at any horizontal level in accordance with the biological species concept of Dobzhansky and Mayr, but he also recognized problems with this concept, which he seemed to think would dissolve away once the horizontal dimension is extended into the vertical dimension as the defining dimension of species (52–53; cf. Stamos 2003, 256–257). The *vestige* here is the *relational* nature of species according to the biological species concept, such that a species needs other species in order to exist, reproductive isolation being meaningless if there are no other species to be reproductively isolated from. Hence Hennig defined speciation in terms of a "cleavage" (58, 64, 66), the first cleavage inaugurating the beginning of a species' existence, the second inaugurating its end (Hennigian extinction), so that species are delimited "by two successive processes of speciation" (63).

What follows from Darwin's species concept, however, in particular his rejection of sterility and fertility, is the sensible view that speciation is coupled to evolution, in other words, that both anagenesis and cladogenesis are speciation processes. Why is this sensible? It is sensible because it avoids a major prejudice, a prejudice that is common to modern phylogenetic systematics. The prejudice can be uncovered as follows. Viewed from the horizontal dimension, most will agree that a species can evolve from being monotypic (having no subspecies or varieties) to being polytypic (having subspecies or varieties), all the while retaining, as a species, its numerical identity. In evolving thus, most will also agree that it has undergone *part* of the process of speciation, namely, cladogenesis. But if the same monotypic species should evolve by the same amount (*same* in the sense of any of its subspecies or varieties in the first scenario) while still remaining monotypic, then why should this not also be considered part of the speciation process, in this case, anagenesis? To allow that the change in the first scenario is speciation but not the change in the second scenario seems to me nothing more than plain and simple prejudice, a prejudice that has its roots, I suspect, in the *relational* aspect of the biological species concept, and that whatever its roots is contrary to the deepest insights of Darwin, in particular his language analogy for species. Of course once the above prejudice is removed, as it is in Darwin's species concept, speciation is necessarily coupled to evolution and the horizontal dimension for species ontology takes its rightful place of priority.

Or rather for Darwin speciation is coupled to *adaptive* evolution. And here one might argue reside the main flaws in Darwin's species concept. For a start, it is questionable whether natural selection is a genuine law of nature (Stamos 2006). But more importantly there is an epistemological problem with Darwin's species concept, in that the determination of which characters are adaptive and which are not is far from an easy task, such that it is often subjective and subject to change (cf. Stevens 1980, 344–345). For example, one has to wonder if, on the topic of whether humanity is one species or more (cf. chapter 5), Darwin would change to a polygenist once he learned (as modern biologists know) that human skin color is not a neutral character but an adaptive character. Would humans then become like primroses and cowslips? And how many today would accept *that*?

A further problem concerns the topic of speciation. By restricting species to adaptations, as I have shown Darwin did, and moreover by restricting adaptations to natural selection, Darwin was resistant to the concept of instantaneous speciation. Darwin, of course, did not know about polyploidy (or genetic engineering if we want to bring that in). The problem is that polyploidy does indeed seem to make, in one generation and commonly in plants, good and distinct species, moreover some that are adaptive and some that are maladaptive. Either way they are good and distinct species, even if the latter are quickly driven to extinction. A viable modern species concept cannot afford to ignore any of this. It therefore should not take natural selection to be the only cause of speciation. Darwin granted to other causal processes a role in species modification, but as we have seen in this chapter he allowed none of them an exclusive role in the formation of adaptations. My suggestion is that Darwin's species concept needs to be weaned from its emphasis on adaptations and natural selection, it needs to be weaned from one *vera causa* to many, so that the similarity complexes that are species are bounded both by the horizontal dimension on the one hand and by a variety of causal mechanisms (natural selection included) on the other hand. This I have attempted to do in the final chapter of my book on the modern species problem (Stamos 2003, ch. 5).

The above suggestion returns us, however, to a further virtue of Darwin's species concept. It concerns the issue of whether a universal species concept is today a viable idea anymore. Many biologists have become skeptical, embracing the idea of species pluralism, that one species concept cannot possibly satisfy all of modern biology, so that modern biology needs a variety of species concepts (e.g., Mishler and Donoghue 1982; Endler 1989, 625; Ehrlich 2002, 347–348 n. 21). Other biologists, however, have expressed not only the hope but the need for a universal species concept. Melissa Luckow (1995), for example, points out that "a universal and definitive species concept is most important to those systematists who actually circumscribe and name species" (589), to which she adds "the species problem will be solved by the continued collection and analysis of data, the clarification of issues and terms, and the application of new ideas" (600). Joel Cracraft (1987) adds that until this happens the Modern Synthesis is incomplete, since "the 'synthetic theory of evolution' has incorporated very disparate concepts of species, so much so that it undermines the very notion of a 'synthesis'" (334).

How does Darwin help? First, with the help of his language analogy for species, we have seen in chapter 3 that for Darwin the primary reality of species is horizontal. Darwin could not have known this, but one consequence of this move is that it does some serious damage to species pluralism, since contrary to species pluralism the priority of the horizontal dimension for species reality privileges some species concepts over others. The result is not that we are left with one species concept, but the door is certainly opened to that possibility (cf. Stamos 2002, 192, 2003, 96–97).

The second way Darwin helps is by emphasizing pattern over process. This is a further distinction by which modern species concepts can be dichotomously classified. Process species concepts view the causal processes responsible for the existence of particular species as part of the ontology of species, whereas pattern species concepts view them apart from the ontology of species. The biological species concept, for example, is a process species concept, while the morphological species concept is a pattern species concept. Darwin's species concept is also a pattern species concept, since the process of natural selection is not for him part of what a species is. The virtues of pattern species concepts over process species concepts are many. One is that epistemologically they are far more accessible, since they focus only on the product, not on the causal processes responsible for the product (Cracraft 1989, 34), and the causal processes responsible for the product are largely untestable (Luckow 1995, 591–592). Second, pattern supervenes on process, in that similar products can be produced by dissimilar causal processes (Levin 1979, 384). Third, pattern species concepts tend to be neutral as to the causal processes responsible for species and therefore do not stand or fall with those theories, which is a virtue since biologists already know that the causal processes are many and may still be more complicated than they already think, and also because the causal processes may well be different in different species (Cracraft 1983, 102, 1989, 34). Fourth and finally, it is only in focusing on pattern rather than on process, along with the priority of the horizontal dimension, that hope for a universal species concept resides. As Luckow (1995) put it, "if species are viewed as the endpoints or results of various processes rather than as participants, a universal species concept is still viable" (590).

Darwin's species concept, of course, does not quite instantiate the above virtues, since he did not cleanly separate the product, adaptations, from the process, natural selection (as we have seen in Chapter 1 on the topic of continuous natural selection). Nevertheless, in developing and employing what was clearly a pattern species concept, Darwin made that first important step and once again helped show us the way.

Chapter 7

The Varieties Problem

Given, as we have seen, that Darwin thought species are real, also that he thought varieties are incipient species, the species problem then becomes, when studying Darwin, the varieties problem. If species are real and varieties are incipient species, then varieties would have to in some sense be real too. In this chapter we shall compare Darwin's view on varieties with those of his contemporaries. The latter topic especially is important as a preparation for the topic of the next chapter, the question of why Darwin would repeatedly define species (both taxa and category) as not real and yet treat them as real, and not only as real but moreover repeatedly and consistently employ his own implicit species concept at that. In other words, to understand Darwin's strategy, we have to understand both Darwin and his contemporaries on the nature of varieties.

A good place to begin is with two passages examined in chapter 1, in which Darwin compares different species concepts and comments on the issue. In a letter to Hooker (December 24, 1856) Darwin wrote

> I have just been comparing definitions of species . . . . It is really laughable to see what different ideas are prominent in various naturalists minds, when they speak of "species" in some resemblance is everything & descent of little weight—in some resemblance seems to go for nothing & Creation the reigning idea—in some descent the key—in some sterility an unfailing test, with others not worth a farthing. It all comes, I believe, from trying to define the undefinable. [Burkhardt and Smith 1990, 309]

131

The analysis presented in chapters 2–6 should allow us to say what Darwin really thought about each of these criteria. Although resemblance was not everything, it was not nothing either, because of common descent. Moreover descent for Darwin was certainly not of little weight but was instead of great weight, because only through descent for Darwin were adaptations acquired, and yet common descent was not the key, as we have seen in the case of primroses and cowslips. As for sterility, it was not for Darwin an unfailing test, nor was it not worth a farthing, but instead it was a fairly reliable test of species distinctness, although it was no part of what a species is. Finally, from Darwin's practice we can see that he did in fact have an implicit definition of "species."

The passage quoted above, therefore, is highly disingenuous. What makes this even more apparent is that Darwin does not mention Edward Blyth's criterion, which we have seen in Chapter 4 on the topic of domestic cattle, and which we have seen in the previous chapter has proven to be of great importance to Darwin, namely, constant and distinct characters. In the next chapter we shall see how the above disingenuous passage fits with what I shall argue was Darwin's strategy. For the present, it is useful to compare it with a highly similar passage from *Natural Selection*, found near the beginning of Chapter IV, the first draft of which Stauffer (1975, 92) tells us was written, according to Darwin's Pocket Diary, from mid-December 1856 to late January 1857:

> . . . how various are the ideas, that enter into the minds of naturalists when speaking of species. With some, resemblance is the reigning idea & descent goes for little; with others descent is the infallible criterion; with others resemblance goes for almost nothing, & Creation is everything; with others sterility in crossed forms is an unfailing test, whilst with others it is regarded of no value. At the end of this chapter, it will be seen that according to the views, which we have to discuss in this volume, it is no wonder that there should be difficulty in defining the difference between a species & a variety;—there being no essential, only an arbitrary difference. [Stauffer 1975, 98]

The similarity of this passage to the other is remarkable, and indeed it is possible that they were written on the very same day. What is especially significant for our present purpose, however, is the difference in the endings. For unlike the previous passage, Darwin in the above passage relates the issue to the difference between species and varieties, stating that there is no essential difference between them. The same claim would resurface in the *Origin*, where Darwin says that in his view "the term species . . . does not essentially differ from the term variety" (52), and again that "there is no essential distinction between species and varieties" (272). The same claim also finds echoes in his correspondence, for example, in his letter to Lyell (September 28, 1860) in which he wrote "Is not your feeling a remnant of that deeply-impressed one on all our minds, that a species is an entity,—something quite distinct from variety?" (Burkhardt *et al.* 1993, 397).

All of the above raises the conceptual issue of what Darwin meant when he said there is no *essential* distinction between a species and a variety. According to Ghiselin (1969), Darwin "insisted on Aristotelian definition as a criterion of reality or naturalness" (82), so that Darwin "maintained that there are no 'essential' differences between species and varieties, and that both terms designate the same basic kind of entity." Thus, "Darwin was denying the reality, not of taxa, but of categories" (93). Similarly, according to Beatty (1985), "the crucial issue was that there was no way of defining 'species' that distinguished species from varieties—no way of defining the difference" (275). More specifically, "*what are called 'varieties'* of species are, in time, transmuted into *what would be called 'species'* in their own right" (276). Thus, "Darwin's theory of the evolution of species was, of course, about the evolution of species taxa rather than the evolution of the species category. So his denial of the species category did not render his theory domainless" (278).

As we have seen in chapter 2, Ghiselin's claim that Darwin insisted on Aristotelian essentialism for naturalness simply does not hold up to close scrutiny, and Beatty's claim that Darwin simply took what his fellow naturalists called "variety" and "species," as we have seen in chapter 5, fares no better. We shall therefore have to take a completely different approach to the topic of Darwin on varieties.

To cut to the chase, it seems to me that Darwin, in claiming that there is no essential difference between a species and a variety, was not claiming that neither is real, nor was he claiming that they are the same basic kind of entity. We shall see later in this chapter, based on the reconstruction of Darwin's species concept developed in previous chapters, that Darwin thought that varieties too are real, only that their reality is partially different than that of species. But equally important, in claiming that there is no *essential* distinction between them, Darwin was merely claiming that there is no *bridgeless gap* between them. That is what essentialism meant to Darwin (and to many today). That is what Darwin meant (at least in part) when he wrote to Lyell, quoted above, that a species is not a *distinct entity*. Again, if species for Darwin are real (and we have seen that they are), and if varieties are incipient species (unquestionably Darwin's view), then it would have to follow that varieties too are real, *but it would not have to follow that they are real in the exact same sense as species.* And indeed this is what we shall find in this chapter. What we have found in chapter 5 is that for Darwin the key feature of species is unique adaptations created and fixed by natural selection. What we shall find in the present chapter, however, is not that there is a distinct difference between species and varieties, but instead that there is a *gradual* difference, so that at one point we can say that this is clearly a variety and at another point that this is clearly a species, but that also somewhere in between we cannot say either way. What we shall see, then, is that not only, once again, did Darwin not insist on Aristotelian essentialism as his criterion of reality, but also that he implicitly denied Aristotle's guiding principle (essential to his essentialism) that whatever is different in degree is not different in kind (e.g., *Parts of Animals* 644ª16–18, *Politics* 1259ᵇ36–38).

What adds a wrinkle to this view, however, is that in the *Origin* Darwin tells us that "every one admits that there are at least individual differences in species under nature. But, besides such differences, all naturalists have admitted the existence of varieties, which they think sufficiently distinct to be worthy of record in systematic works" (468–469). The claim is also found earlier in *Natural Selection*, but with some interesting twists: "this admission [subspecies by authors 'of the highest authority'—Darwin cites six] is important as sub-species fill up the gap, between species, admitted by everyone & varieties admitted by everyone. Between varieties & individual differences there seems a gradual passage but to this subject we shall recur" (Stauffer 1975, 99). Darwin's motive behind these assertions is plain enough, and it becomes even more obvious at the bottom of the paragraph from which the above passage is taken: "So that between individual differences & undoubted species naturalists have made various short steps" (99).

What is plainly clear is that Darwin *wanted* to believe, or more precisely wanted his readers to believe, that his fellow naturalists believed in the reality not only of variations but also of varieties and subspecies and in fact every grade between individual variation and species, the reason being one more piece of evidence in his one long argument for evolution by natural selection, in particular that varieties are "incipient species." As Darwin put it in the *Origin*, "These differences blend into each other in an insensible series; and a series impresses the mind with the idea of an actual passage" (51). Moreover it explains why Darwin would repeatedly focus, in both *Natural Selection* (Stauffer 1975, e.g., 112–113, 159–164) and the *Origin* (e.g., 48, 404), on the many doubtful forms called "species" by some and "varieties" by others, a point to which he owed much to Watson (cf. letter from Watson, November 19, 1856, Burkhardt and Smith 1990, 276–277).

But did Darwin's fellow naturalists actually believe in the existence of varieties as Darwin would have us believe? This is a question that remains to be answered. In what follows I shall argue that the answer in the main is no, and moreover that Darwin knew it. Darwin's knowledge that he was misleading will prove to be especially significant when we turn to his strategy examined in the next chapter.

The first point of evidence is that although naturalists gave plenty of definitions of the concept of species (the commonality was creationism, as we have seen in chapter 1), they said relatively little with regard to the nature of varieties. Consequently, unlike species, we have to really work at figuring out what their concept of variety was.

Perhaps the best place to begin is with those who explicitly denied that varieties exist. Louis Agassiz is a good case in point. In an often-cited passage, Agassiz (1860a) claimed that "varieties properly so called, have no existence, at least in the animal kingdom" (410). It might be replied that this was simply a rearguard move on Agassiz's part in response to the *Origin*.[1] Mayr (1976, 258, 259), however, argues that this was not the case, that this was a view that Agassiz held well before and in spite of the *Origin*. Mayr points out that of course Agassiz recognized the existence of individual variations—who didn't?—but he did not recognize the existence of varieties as subspecific taxa. Agassiz avoided ascribing reality to varieties by applying two different approaches. One was to be a splitter and to make a species out of every definable distinction. Mayr points out that Agassiz did this in the case of his study of the fish of the Tennessee River, published in 1854. As Mayr puts it, "His disbelief in the existence of 'varieties' forced him to describe several 'species' from schools of single species" (258). Agassiz's other approach was to unite into one species forms that insensibly intergrade, as he did in the case of echinoderms (258–259). In all of this we can see his rejection of varieties as being consistent with his Platonic/Christian approach to species. For Agassiz, as he stated in his *Essay on Classification* (1857), "the individuals living at the time have alone a material existence," while "branch, class, order, family, genus, and species . . . . have been instituted by the Divine Intelligence as the categories of his mode of thinking" (8). Again: "individuals . . . do not constitute the species, they represent it. The species is an ideal entity, as much as the genus, the family, the order, the class, or the type" (175–176). Moreover no one individual "exhibits at one time all the characteristics of the species" (176). Each species, then, for Agassiz, is in reality only an idea in the mind of

God, not an entity in nature. This comes out in various places in his *Essay*; it also comes out clearly and succinctly in his review of the *Origin* (1860b). There he says "while individuals alone have a material existence, species, genera, families, orders, classes, and branches of the animal kingdom exist only as categories of thought in the Supreme Intelligence, but as such have as truly an independent existence and are as unvarying as thought itself after it has once been expressed" (151). To this he adds that each species has "essential" or "specific characters" which "are forever fixed," such that "*individual peculiarities*" are "in no way connected with the essential features of the species, and are therefore as transient as the individuals" (150).[2]

The case of Agassiz raises an issue that Darwin was certainly aware of, namely, that most naturalists from the time of Linneaus to his own day were splitters, not lumpers. The evidence for this is overwhelming. Carolus Linnaeus, for example, the self-proclaimed Prince of Botanists, complained of the varieties "which the common herd of botanists calls species," and for him such splitting is wrong because the variation is merely "in the outside shell" (Ramsbottom 1938, 198 n.). Jean Baptiste de Lamarck, in his *Flore Francoise*, published in 1779 and before he became an evolutionist, complained that "nearly all present-day botanists" are splitters, "multiplying the species, at the expense of their varieties, to infinity" (Burkhardt 1987, 163). H.C. Watson (1845b) in England wrote of the "species-splitting *monomania*" (219), while Hooker and Thomson in their *Flora Indica*, published in 1855, complained that "the study of systematic botany is gradually taking a lower and lower place in our schools; it falls into the hands of a class of naturalists, whose ideas seldom rise above species, and who, by what has well been called *hair-splitting*, tend to bring the study into disrepute" (Stevens 1997, 354). Darwin himself notes in his *Origin* that "differences, however slight, between any two forms, if not blended by intermediate gradations, are looked at by most naturalists as sufficient to raise both forms to the rank of species" (485; cf. 424–425). Indeed in a letter to Leonard Jenyns (April 28, 1858) Darwin complained, in line with the complaints above, that "One chief reason why I have not accumulated more facts of variation in a state of nature is, that naturalists so invariably turn round & say oh they are not varieties, but species" (Burkhardt and Smith 1991, 86). Two years later in a letter to Lyell (June 14, 1860) he would make basically the same complaint, this time specifically about Agassiz: "It is no wonder Agassiz denies varieties in animals, when he calls even the *same* forms in two distant countries, two species" (Burkhardt *et al.* 1993, 254). And yet "all naturalists have admitted the existence of varieties"!

The fact is, or at least I shall argue it, most naturalists did not admit the existence of varieties, in spite of their linguistic habits. A few were explicit about it, but most left it implicit. Certainly it seemed to follow from their essentialism. A good case in point is Linnaeus. Confining ourselves to his earlier works, for which he was most famous, Linnaeus states, in for instance his *Fundamenta Botanica*, published in 1736, that "There are as many varieties as there are different plants produced from the seed of one and the same species" (Leikola 1987, 49)—"Varietates tot sunt, quot differentes plantae ex ejusdem speciei semine sunt productae" (Ramsbottom 1938, 199). This might be a simple case of what Blyth (1835) wrote about in his attempt to classify the different ways in which naturalists used the term "variety." According to Blyth, "The term 'variety' is understood to signify a departure from the acknowledged type of a species, either in structure, in size, or in colour; but is vague in the degree of being alike used to denote the slightest individual variation, and the most dissimilar breeds which have originated from one common stock" (40–41). On the other hand, Linnaeus' definition above might be interpreted as meaning to say that there are as many varieties of a species as there are individuals of that species with variations. In other words, it is a cynical treatment of varieties. Variations are character states that individual organisms have, whereas varieties are subspecific taxa made up of one or more individual organisms. Linnaeus, then, it seems to me, is saying that varieties are arbitrary classifications. And indeed for Linnaeus, raised in one of the last strongholds of Aristotelian scholasticism in Europe, the University of Uppsala in Sweden (Leikola 1987, 45–47), species are essentialistic kinds, created from the beginning and remaining essentially unchanged (49, 56 n. 10), defined binomially by a genus and differentia (which was a shorthand for a full description of essential characters), genus level characters being generative organs, species level characters being nutritive parts such as roots, stems, and leaves (49, 56 n. 10). Variety characters, on the other hand, were based on characters that resulted from accidental causes such as soil and climate—"Varietas est Planta mutata a caussa accidentali: Climate, Solo, Calore, Ventis, &c., reducitur itaque in Solo mutato" (Ramsbottom 1938, 199)—and also culture—"Naturae opus semper est Species & Genus; culturae saepius Varietas" (199–200)—characters such as, in the case of plants, size, color, smell, and taste—"Species varietatum sunt Magnitudo, Plenitudo, Crispatio, Color, Sapor, Odor" (Ramsbottom 1938, 199). But equally important, Linnaeus confined varieties, or more specifically the variations on which they were based, not to the deep essence of the species, something that was passed on in the "unity of generation" (Leikola 1987, 50), as he called it in

Systema Naturae, published in 1735, but rather, as we have seen above, to what he called in *Critica Botanica*, published in 1737, "the outside shell" (Ramsbottom 1938, 198 n.).

For all of the above reasons, and more, I would agree with Larson's (1968) view that for Linnaeus varieties were not real (291). As we have seen in chapter 2, for Linnaeus categories were conceived to be like boxes (Leikola 1987, 46). But Linnaean varieties do not at all fit this scheme. They have no underlying essence, there are no bridgeless gaps between them (recall from chapter 5 that Linnaeus was a lumper), they are based on characters that not only are not fixed but are infinite in their possibilities (this comes out from his cynical definition of "variety" quoted above), and finally, Linnaeus believed in the so-called law of reversion, the belief that species can vary within certain "fixed limits" and that when the conditions that caused them to vary are removed, "true species," as he put it in *Critica Botanica*, will "finally revert to the original forms" (Ramsbottom 1938, 200 n.). In all of this it is very difficult to see Linnaeus as falling under Darwin's claim, quoted above, that "all naturalists have admitted the existence of varieties."

We can see the same in Linnaeus' contemporary and nemesis George Louis Buffon, highly influential in France and the first big name among French naturalists. Although Buffon expressed species nominalism at the beginning of his *Histoire Naturelle*, published in forty-four volumes between 1749 and 1804, he quickly took up species essentialism, arguing that each species has what he called an internal mold (*moule intérieur*), which causally preserved the form of the species from generation to generation. Of this mold he stated that "The type of each species is cast in a mold of which the principle features are ineffaceable and forever permanent, while all the accessory touches vary" (Stamos 1998, 449). Granted, Buffon also developed a theory of degeneration, extending it more and more as his writings progressed. According to this theory many species have degenerated over the ages from a common stem (*souche première*), resulting in what are wrongly thought to be species, such as the ass and zebra, which for Buffon are merely varieties of the horse (447–448). Hence Buffon's species are much more inclusive, more like Linnaeus' genera. But even so, we can see in this what is arguably only a much more extreme view of the plasticity of species that we find in Linnaeus and other essentialists before Darwin. For Buffon these varieties do not have their own internal mold, each with its essential characters, but are merely incremental perturbations, "accessory touches" that "vary," often cumulative and heritable but always reversible, produced by differences in environmental conditions and domestication (Farber 1972, 268–278).

The idea that variations in a species are perturbations from the norm or essence of the species had quite a wide currency. In addition to Linnaeus and Buffon, another enormously influential example is Antoine-Laurent de Jussieu, whose natural system (which involved taking into account the entire organization of each individual, not just characters thought to be privileged for classification) served as the basis for plant systematics for well over a century, and not just in France but throughout continental Europe and England (Stevens 1994, xx). In the Introduction to his *Genera Plantarum*, published in 1789, Jussieu defined species as

> a collection of beings that are alike in the highest degree, never to be divided, but simple by unanimous consent and simple by the first and clearest law of Nature, which decrees that *in one species are to be assembled all vegetative beings or individuals that are alike in the highest degree in all their parts, and that are always similar over a continued series of generations*, so that any individual whatever is the true image of the whole species, past, present, and future. [Stevens 1994, 356–357]

This is a clear example of an essentialist species concept. Indeed a little earlier in the same work Jussieu states that species are "immutable and perpetual in kind" (340). And yet he allowed for varieties, but in a way that, like Linnaeus, hardly made them real. This is because he allowed species to be "occasionally subverted for a while by chance or human industry" (340–341), to vary in characters such as color or size of organs. Moreover he states that "these *varieties*, obeying the law of nature, and committed to a new germination of seed; they return to the primordial species, their character restored, if other factors do not interfere" (341). Not only do we find in this the law of reversion, but again, just like Linnaeus, we find the view that variations are perturbations from the norm, as well as the logical consequence (though it is implicit) that naming varieties (subtaxa based on these variations) is ultimately arbitrary.

I will return to the law of reversion shortly. For the present I want to continue with the idea of variations as perturbations. Among influential sources, especially in England, James Prichard (1813) is a good example. Prichard subscribed to the common view (as we have seen in chapter 1) that all classificatory categories are matters of "arbitrary definition," the exception being the species category, where "the distinction is formed by nature" (7). Nevertheless he also states that "it is well known that considerable varieties arise within the limits of one species, and such varieties often become to a great degree hereditary in the race, and permanent" (8).[3] To determine whether a particular group is part of a real species "distinct from their first creation" (8) or part of a variety (which developed from a real species), Prichard referred to what he called "indirect methods of reasoning" (8), which for him had to be the sterility test (12–13). What is interesting is that he then goes on to discuss "the kinds of variation in which Nature chiefly delights" (13), and it is clear that like so many others he confused varieties with variations. This comes out clearly when he states

> When we have found that any particular deviation from the primitive character has taken place in a number of examples, the tendency to such variety may be laid down as a law more or less general, and accordingly when parallel diversities are observed in instances, which do not afford us a view of the origin and progress of the change, we may nevertheless venture to refer the latter with a sufficient degree of probability to the class of natural varieties, or to consider them as examples of diversified appearance in the same individual species. [13–14]

Prichard then goes on to give the example of albinoism in mice, rats, and crows (14), and further states that the most common kinds of change are "varieties of form and colour" (15). We then see what one finds in the literature of this period time and again, namely, the idea of variations, including heritable variations, as perturbations from the norm:

> We may however in general observe, that when the condition of each species is uniform and does not differ materially from the natural and original state, the appearances are more constant, and the phænomena of variation, if they in any degree display themselves, are more rare and less conspicuous, than when the race has either been brought by human art into a state of cultivation, or domestication, or has been thrown casually into circumstances very different from the simple and primary condition. [15–16]

What only underscores Prichard's confusion of variations with varieties is when he applies his principles to the human species (Prichard was a monogenist), naming as varieties the albino (17–19), red or yellow hair with fair skin (20), black hair and dark eyes with white but tanable skin (21–22), yellow to olive skin with long black hair ("the Mongoles," etc.) (22–23), "Native Americans" (23), "pie-bald Negroes" (blacks with patches of black and white skin and hair) (23–24), and finally black or dark yellowish-brown skin (24). What is interesting about Prichard is that he does not seem to have subscribed to the law of reversion, which as we shall shortly see was generally accepted as a natural law in England and elsewhere.

Like Prichard, Edward Blyth (1835) also allowed for permanent varieties, but he was more circumspect on how his fellow naturalists used the term "variety" and also on the law of reversion. Dividing those uses into four general categories, the first, as we have seen, being the confusion with individual variation, Blyth's second category is acquired variations either in a single individual or in a series of individuals in which "various changes . . . are *gradually* brought about by the operation of known causes" (43), while his third category is breeds, typically produced by human agency but he allows that they could be produced in nature as a result of "accidental isolation" (45). The interesting thing about breeds, according to Blyth, is that "if man did not keep up these breeds by regulating the sexual intercourse, they would all naturally soon revert to the original type" (46). Blyth's fourth and final category, interestingly the only one to which he thinks the term "variety" properly applies, and he knows this restriction is "peculiar" (47), is what he calls "*true varieties*," and they are what Prichard and others would call "permanent" varieties. These true varieties, he says, "are, in fact, a kind of deformities, or monstrous births, the peculiarities of which, from reasons already mentioned, would very rarely, if ever, be perpetuated in a state of nature; but which, by man's agency, often become the origin of a new race" (47). Some of the examples he gives are ancon sheep, with their legs too short to jump over fences, tailless cats, many kinds of domestic dogs, and fan-tailed pigeons. Of true varieties, he says, "The deviations of this kind do not appear to have any tendency to revert to their original form" (47).

What follows from Blyth's analysis is that what are called "varieties" in nature are probably not such at all, that varieties in nature probably do not exist. Interestingly, for further reading Blyth strongly recommends Prichard (1813), to which he adds "some sound and excellent remarks on *varieties* will also be found in the second volume of Lyell's *Principles of Geology*" (48).

When we turn to Lyell (1832), however, we find ideas that do not seem congenial to Blyth's analysis. In chapter 1 we have already seen that Lyell, like so many others, held the premise that if species are not fixed but evolve, then species are not real. It would be odd, then, to the point of inconsistency, if Lyell at the same time thought that varieties are real. This is because he did not think they are fixed, any of them. Moreover he seems to have equated varieties merely with variations, including heritable variations. He clearly thought of "varieties" as "the slight deviation from a common standard of which a species is capable" (62), and he allowed for specific characters to vary as well (25). He allows that "the organization of individuals [of a species] is capable of being modified to a limited extent by the force of external causes" and that "these modifications are, to a certain extent, transmissible to their offspring," but he stresses that species have "fixed limits" (23). Interestingly, Lyell thought of the "common standard" of a species, along with its complete possible range of variation, as belonging to the ontology of each species: "the mutations thus superinduced are governed by constant laws, and the capability of so varying forms part of the permanent specific character" (64). On such a view it would therefore be impossible to say how many varieties a particular species has, even if its full range of variation were expressed, unless one employed the cynical definition of "variety" that we have seen in Linnaeus. To all of this, Lyell also added his vote to the "law of reversion" (28, 33), a law that I shall discuss more fully below, to which he does not seem to have allowed for any exceptions. Moreover, he does not seem to have allowed that what he called "extraordinary varieties" could ever arise in nature and let alone be permanent, for in the case of the "extraordinary varieties" produced by horticulturalists he says they "could seldom arise, and could never be perpetuated in a wild state for many generations, under any imaginable combination of accidents" (32).

Many of these ideas can also be found in Agassiz's colleague in America, the botanist James Dwight Dana. Dana (1857) thought of each species as having "*a specific amount of concentered force, defined in the act or law of creation*" (486), to which he added that each member organism of a species has "the specific law of force, alike in all" (487), a clear expression of species essentialism. As for varieties, Dana argued that "there are variations in species" and "variations have their limits, and cannot extend to the obliteration of the fundamental characteristics of a species" (492). Again Dana thought of these as perturbations from the norm caused by outside forces: "There is then a fixed normal condition or value, and around it librations take place. There is a central or intrinsic law which prevents a species from being drawn off to its destruction by any external agency, while subject to greater or less variations under extrinsic forces" (493). And indeed like Lyell above, Dana conceived of the ontology of each species as constituting not only its specific characters but also its full range of possible variation, the "process of variation" being "the external revealing the internal" (493), such that "The many like individuals that are conspecific do not properly constitute the species, but each is an expression of the species in its potentiality under some one phase of its variables; and to understand a species, we must know its law through all its cycle of growth [i.e., its ontogenetic stages, cf. 487–488], and its complete series of librations" (494). In all of this it seems impossible to find the view that varieties are real. Instead his focus is on variations and their ranges. Again he says "For while a species has its constants, it also has its variables" and "The variables are a necessary complement to the invariables; and the complete species-idea is present to the mind, only when the image in view is seen to be ever changing along the lines of variables and development." Thus he says "Whatever individualized conception is entertained, it is evidently a conception of the species in one of its phases,—that is, under some one specific condition as to size, form, colour, constitution, &c., as regards each part in the structure, from among the many variations in all these respects that are possible" (495). On such a view, the language of varieties might be helpful, and indeed necessary for discourse, but the varieties themselves would not be real since they are picked from ranges of many variables.

One attempt to circumvent this kind of conclusion is to be found in Thomas Vernon Wollaston (1856), who confined himself to his specialty, insects. Acknowledging that it has "frequently been asserted that everything is to be regarded as a 'variety' which has wandered in the smallest degree from its normal state," Wallace contends that this is "essentially an error" and asserts in its place that for a variety to be real it "must have in it the *prima-facie* element of stability." Wollaston recognizes that such a variety might well be taken for a species, and for him "even small differences *should be regarded as specific ones* so long as the intermediate links have not been detected which may enable us to refer them to their nearest types" (6). Consequently he provides the additional criterion that a variety must have "intermediate links (which, although rarer than the variety itself, *must nevertheless exist*) to connect it with its parent stock" (5). For Wollaston, then, not only must a variety be stable and connected by intermediate links with its parental stock, but it must also have greater numbers of organisms than the intermediate links in order to be real. This is an odd foundation on which to base the reality of varieties, but it is exactly what Wollaston did. He gives a general example found in many darkly colored insects, namely, that they "vary, by slow and regular gradations, into a pallid hue, sometimes into almost white." For Wollaston, if "the *extreme* aberration" has greater numbers than the intermediate ones, then "there is but a *single* variety involved, namely a pale one,—the gradually progressive shades which imperceptibly affiliate it with its type not being regarded in themselves as 'varieties' at all" (5). Otherwise, says Wollaston, one would have to regard as a variety "*every* separate degree of colour which could possibly be found between the outer limits" (6), and that was something Wollaston was not prepared to do. Between the two extremes, though, Wollaston was prepared to admit more than one variety, if, which he says sometimes occurs, "between the two extremes, there are several nuclei, or centres of radiation, to which the names of varieties may be legitimately applied" (6).

Darwin, interestingly, found Wollaston's suggestions intriguing, even having "a good deal of truth" and being "very important" (Burkhardt and Smith 1990, 171), which should not be surprising given that they lend themselves to Darwin's principles of natural selection and divergence. He consequently in the same year (1856) queried Hooker, Gray, and Watson in his correspondence to get further facts and corroboration (171, 182, 208). Hooker (July 10) replied that the rare intermediates can always be explained away by hybridism (176). Gray (early August) replied with his impression that the rarity of intermediates is often true but that Wollaston's approach went against the naming of a species by type specimen, since the type specimen often turned out to be the rarer form (195). Watson (December 25) replied similarly that the rarity of intermediates might be shown by many examples, but he added that it makes the distinction between species and variety arbitrary since it entails that whatever form is the more numerous is the species, not the variety, and this would have to be switched if the numbers accordingly changed (310)—a point he had made years earlier to Hooker (Burkhardt and Smith 1988, 31). Darwin took all of this with a grain of salt, and in *Natural Selection* used the rarity of intermediates as further though minor evidence for gradual evolution, and he attempted to explain why on his principles they would be expected not to last (Stauffer 1975, 268–270; cf. Darwin 1859, 176–177).

Wollaston's position on varieties, of course, was not the norm, and the evidence for this comes from the widespread belief in the so-called law of reversion, which should only make us further skeptical that most naturalists from the time of Linnaeus to Darwin believed in the objective reality of varieties, as a real subcategory distinct from variations. We have seen belief in reversion explicitly in Linnaeus, Jussieu, and Lyell above, while it is partially advocated by Blyth and is implicit in Dana. Darwin himself often refers to this law as "well-known" (e.g., 1859, 25), and Wallace (1859) tells us that it had "great weight with naturalists, and has led to a very general and somewhat prejudiced belief in the stability of species" (10). Wallace describes the law as follows: "*varieties* produced in a state of domesticity are more or less unstable, and often have a tendency, if left to themselves, to return to the normal form of the parent species; and this instability is considered to be a distinctive peculiarity of all varieties, even those occurring among wild animals in a state of nature, and to constitute a provision for preserving unchanged the originally created distinct species." He then adds, with regard to the difficulty the existence of permanent or true varieties posed for these naturalists, that "the difficulty is overcome by assuming that such varieties have strict limits, and can never again vary further from the original type, although they may return to it, which, from the analogy of the domesticated animals, is considered to be highly probable, if not certainly proved" (10–11). Of course for Darwin and Wallace the law of reversion is not a true law. But what is important for our purposes is what it does to the supposed reality of varieties, as a legitimate subcategory of species (which as such would have to have one or more individual organisms) distinct from the category of variation (which is something an individual organism has). It would seem to make the subcategory of variety a mere category of linguistic convenience and nothing more.

This view becomes only strengthened when we examine what naturalists commonly thought was the immediate cause of reversion, the explanation of reversion. This was the *underlying causal essence* of each species (cf. Ellegård 1958, 209). Aristotle seemed to have this idea in mind (cf. Stamos 2003, 106–107), and we have seen evidence of it above in Linnaeus, Buffon, Jussieu, Agassiz, and Dana, though among them there is of course variation in the concept. I suggest that these authors were not the exception but that the concept of each species having an underlying essence was common in the minds of naturalists from the time of Linnaeus to Darwin, both for lumpers and splitters, and not simply as a matter of tradition going back to Aristotle (or even Plato in the case of Agassiz), but more importantly because of the common belief in, supported in part by widespread evidence for, the so-called law of reversion.

All of this once again does damage to the view that naturalists of this period believed in the existence or reality of varieties. Instead it points toward the view that they used the word "variety" as a mere linguistic convenience, for the sake of communication. What further supports this view are the adjectives they used when referring to varieties. Lyell (1832), for example, writes of "mere varieties" (10, 23) and "a mere temporary and fortuitous variety" (51). Even Huxley (1859c), in one of his reviews of the *Origin*, referred to varieties as "accidental," "fleeting," and "mere" (8).

In fact Darwin himself used such terms when referring to varieties, in particular in his correspondence with other naturalists, whom he well knew did not think much of varieties. For example, in a letter to Asa Gray (February 21, 1858) he wrote "I know what fleeting & trifling things varieties very often are," to which he added "but my query applies to such as have been thought worth marking & recording" (Burkhardt and Smith 1991, 27). Moreover he knew from his main correspondent, J.D. Hooker, that varieties are not important to most naturalists. Hooker himself expressed to Darwin (July 5, 1845) his own view that "mutation" and "Hybridizing &c" are "the perturbing causes of our difficulties in assigning limits to species" (Burkhardt and Smith 1987, 211), but more importantly he once wrote to Darwin (March 14, 1858) that "The long & short of the matter is, that Botanists do not attach that *definite* importance to varieties that you suppose" (Burkhardt and Smith 1991, 49).[4]

This was hardly a revelation to Darwin. Even in his early notebooks we can see him recognize the lack of importance naturalists commonly gave to varieties. For example, in Notebook C (137) he wrote "The simple expression of such a naturalist 'splitting up his species ['& genera' later added] very finely' show how arbitrary & optional operation it is.—show how finely this series is graduated.—Dr Beck doubt if local varieties should be remembered, therefore do not consider it as proved that they are varieties, (though that would be best)" (Barrett *et al.* 1987, 281). Of course Darwin needed varieties to be real so they can be incipient species, hence "that would be best." And yet he also knew what he was up against. Roughly one hundred years earlier Joseph Butler, in his *Analogy of Religion* published in 1736, expressed the prevalent underlying philosophy when he wrote "The only distinct meaning of the word 'natural' is *stated, fixed,* or *settled,*" a quotation Darwin added to the frontispiece of the second edition of the *Origin* (Burkhardt *et al.* 1993, 334 n. 11).

So what then was Darwin's own view on varieties? In previous chapters I have reconstructed his species concept. It remains now to reconstruct his variety concept. Perhaps the first obstacle to overcome is the claim that Darwin, like so many others, confused "variation" with "variety." According to Simpson (1961, 177–178), in many naturalists of the time and beyond it was common to use the term "variety" to denote an individual variant, a group of such variants not forming a population, or a population of such variants distinguishable from the norm of the species. He then goes on to say that Darwin was guilty of this confusion and that it is "probably the most serious logical ambiguity in the usually severely careful work of Darwin" (178). Equally severe is Mayr (1982), who argues that Darwin's early association with botanists caused him to use the word "variety" ambiguously to denote either individual variants or geographic varieties. According to Mayr, "instead of using the term 'variety' consistently for geographic races, he frequently employed it, particularly in his later writings, as a designation for a variant or aberrant individual. By this extension of the meaning of the term 'variety,' Darwin confounded two rather different modes of speciation, geographic and sympatric speciation" (268; cf. 288, 415, and Mayr 1991, 32–33).

Mayr (1982, 496) also makes the same claim about Wallace, that he confounded "variation" and "variety," to which Kottler (1985, 375–378) provides detailed support. In the case of Darwin, however, I just don't see it. For a start, there is his connection with Blyth. We know that Darwin had some considerable correspondence with Blyth in 1855, and also that he thought very highly of Blyth, as he stated in the *Origin* (18). But most of all, Darwin had an annotated copy of Blyth's article on varieties in his library (Burkhardt and Smith 1989, 468 n. 11), so we know he had read Blyth (1835). From those facts alone it would be remarkable if Darwin went on to confuse "variation" with "variety." And indeed when we turn to the *Origin*, we do not see Darwin making this mistake (except perhaps for some minor lapses here and there). Instead he distinguished the two concepts fairly well. For example, when writing on what it is that natural (and by analogy artificial) selection is supposed to work on, Darwin repeatedly says that it is "variations," never that it is "varieties" (e.g., 30, 61, 80, 84, 102, 108, 169, 454–455, 467, 469). Darwin also says that it is "varieties" that are incipient species, never that it is "variations" or "variants" (e.g., 52, 54, 59, 111, 133, 169, 176, 325, 404, 469–470). Moreover near the beginning of this chapter we have seen Darwin make an explicit distinction, in both *Natural Selection* and the *Origin*, between individual differences and varieties. Simpson's claim and Mayr's following it just do not hold up.

As for Darwin on speciation, he was a pluralist, much like many biologists today (including Mayr himself). Speciation by geographic isolation always remained important for him. As he stated in his concluding chapter in the *Origin*, "Widely ranging species vary most, and varieties are often at first local,—both causes rendering the discovery of intermediate links less likely. Local varieties will not spread into other and distant regions until they are considerably modified and improved; and when they do spread, if discovered in a geological formation, they will appear as if suddenly created there, and will be simply classed as new species" (464–465; cf. 301). As for sympatric speciation, we have seen in chapter 3 that Darwin in the *Origin* favored this more than geographic speciation (isolation). But this was not because of any conceptual confusion between "variation" and "variety." Instead it was because of the importance of the role he allotted to his principle of divergence.

Further support for the view that Darwin did not confuse the two concepts becomes evident when we look closely at how he distinguished varieties from species.

To begin, in the concluding chapter of the *Origin* he says "Hereafter we shall be compelled to acknowledge that the only distinction between species and well-marked varieties is, that the latter are known, or believed, to be connected at the present day by intermediate gradations, whereas species were formerly thus connected" (485). In his barnacle work Darwin put this criterion into practice. For example, he says (1854b) "In determining what forms to call varieties, I have followed one common rule; namely, the discovery of such closely allied, intermediate forms, that the application of a specific name to any one step in the series was obviously impossible; or, when such intermediate forms have not actually been found, the knowledge that the differences of structure in question were such as, in *several allied forms*, certainly arose from variation" (156; cf. 197).

But this was not the only difference between species and varieties that we can find in Darwin's mature writings. The fact is, he advocated other criteria as well. Again in the *Origin* he says "Undoubtedly there is one most important point of difference between varieties and species; namely, that the amount of difference between varieties, when compared with each other or with their parent-species, is much less than that between the species of the same genus" (57). The phrase "amount of difference" should ring a bell, for in chapter 5, in particular in the case of primroses and cowslips, we have seen that the intergrading criterion was overruled by the amount of difference criterion. But we also saw in that chapter that in addition to amount of difference was the criterion, for species, of constant and distinct characters, and also that behind that criterion was the criterion of adaptations. This, it turns out, is the real distinction for Darwin between unequivocal varieties and unequivocal species, and it is time to bring it out more clearly.

First, any character upon which a variety is based must for Darwin be heritable. As he puts it in the *Origin*, "Any variation which is not inherited is unimportant for us" (12). Moreover, the variations upon which varieties are based must not be constrained by barriers such as reversion but must in principle be capable of varying to the point of specific difference. Again in the *Origin* Darwin says "It has often been asserted, but the assertion is quite incapable of proof, that the amount of variation under nature is a strictly limited quantity" (468). All of this, which is necessary if varieties are to be "incipient species," is enough to make Darwin's concept of variety remarkably different from those of his contemporaries.

But of course there is more. Darwin distinguished between individual variants and varieties, the difference being that the latter is a subtaxon or the beginning of one, and moreover that natural selection is in operation toward making it a species. This comes out rather clearly in *Natural Selection* (Stauffer 1975), where Darwin distinguishes between species and varieties, the former in his view "being much more constant in all their characters," this constancy being "partly due to the several causes of variability having acted less energetically on the two species under comparison than on the one species yielding the two or more varieties; and partly to the characters of the two species having been long inherited, & by this cause having become more fixed." Hence, he says a few lines later, "as a general rule, species may be looked at as the result of variation at a former period; & varieties, as the result of contemporaneous variation" (165). This is from his chapter IV titled "Variation Under Nature," which comes two chapters before his chapter VI titled "On Natural Selection." When Darwin does get to that chapter, he connects this constancy of characters with natural selection:

> In regard to the difference between varieties and species, I may add that varieties differ from each other & their parents, chiefly in what naturalists call unimportant respects, as size, colour, proportions &c; but species differ from each other in what naturalists consider more important respects. But we have seen in Chapter IV, that varieties do occasionally, though rarely, very [sic, vary] slightly in such important respects; and in so far as differences in important physiological characters generally stand in direct relation to different habits of life, modifications however slight in such characters would be very apt to be picked out by natural selection & so augmented, thus to fit the modified descendants from the same parent to fill as many & as widely different places in nature as possible. [244]

As for the origin of a variety, in particular a variety "in some degree permanent," back in Chapter IV Darwin says "whether it has originated in a single accidental variation, or by the addition of several such successive variations through natural selection, or through the direct & gradual action of external conditions, as of climate, its first origin is even of less importance to it, than its preservation" (137). Its preservation, of course, as well as its modification into a species, will be effected by natural selection. This is what makes varieties incipient species. And naturally Darwin was ready to admit that he was "far from supposing that all varieties become converted into what are called species: extinction may equally well annihilate varieties, as it has so infinitely many species" (159).

In all of this we can see that Darwin had a clear distinction in mind between variation and variety. This comes out, for instance, later in chapter VI:

> It should always, be borne in mind that there is a wide distinction between mere variations & the formation of permanent varieties. Variation is due to the action of external or internal causes on the generative systems, causing the child to be in some respects unlike its parent; & the differences thus produced may be advantageous or disadvantageous to the child. The formation of a permanent variety, implies not only that the

modifications are inherited, but that they are not disadvantageous, generally that they are in some degree advantageous to the variety, otherwise it could not compete with its parent when inhabiting the same area. The formation of a permanent variety must be effected by natural selection; or it may be the result, generally in unimportant respects, of the direct action of peculiar external conditions on all the individuals & their off-spring exposed to such conditions. [240]

All of this was prefigured by Darwin in one of his many letters to Hooker (November 23, 1856), written a few weeks before he began the first draft of chapter IV of *Natural Selection* and roughly five months before he completed the first draft of chapter VI (Stauffer 1975, 213):

> . . . my conclusion is that external conditions do *extremely* little, except in causing mere variability. This mere variability (causing the child *not* closely to resemble its parent) I look at as *very* different from the formation of a marked variety or new species.—(No doubt the variability is governed by laws, some of which I am endeavoring very obscurely to trace).—The formation of a strong variety or species, I look at as almost wholly due to the selection of what may be incorrectly called *chance* variations and variability. [Burkhardt and Smith 1990, 282]

In all of this, we can see not only that Darwin, contra Simpson and Mayr, did not confuse "variation" with "variety," but that he developed a concept of variety that was revolutionarily different from anything that had been used before. By raising the importance of variations, which hitherto were thought to be unimportant, by raising the reality of varieties and redefining them as incipient species, which hitherto were thought to be neither, and by focusing on the production of adaptations by natural selection, Darwin accomplished, or attempted to accomplish, an enormous paradigm shift.

But in saying that between species and varieties there is no essential difference Darwin was *not* saying that species and varieties are not real. Instead he was simply saying that there is no barrier preventing a real variety from becoming a real species. To say that there is an essential difference is to say that there is such a barrier. For species essentialists in Darwin's day, there was such a barrier. As Watson wrote in a letter to Darwin (May 10, 1860), "Until a faith in certain impassable barrier between existent species becomes thoroughly shaken, naturalists will resist your views, & hail difficulties as if conclusive arguments on the contra side. Differently as these unseen barriers are traced or placed, they are believed in about as strongly by almost all" (Burkhardt *et al.* 1993, 203). What Darwin was doing was arguing against the existence of that barrier. In a very important way, then, ironically, his concept of variety, *not* his concept of species, was the Archimedean fulcrum with which he would shake and eventually move the world.

The fact remained, however, that Darwin's contemporary naturalists did not think of varieties, and consequently of species, as he himself did. In order to shake their faith in what Watson called the "certain impassable barrier," more than straight exposition must have seemed to Darwin required. We are thus led to Darwin's strategy.

Chapter 8

Darwin's Strategy

If the previous analysis is basically correct, it raises the obvious question of why Darwin would define "species" nominalistically, both taxa and category. If he was indeed a realist, it seems extremely odd that he would do this, especially since it is duplicitous and he was, after all, a man of honor. In this chapter I will develop and defend a novel answer to this problem. Since Darwin, as far as is known, never provided an explicit answer to this question, and since it is obviously too late to get inside his head in the manner of a Schilpp volume, the best that we can do, once again, is to apply the principle of charity and make the best of the evidence that we can. I believe that the answer I develop has considerable evidence in its support. But first I want to look at the answers given by Ghiselin, Mayr, Beatty, Sulloway, Hodge, McOuat, and my former self. The failings of their answers will further justify the search for a new solution.

Ghiselin (1969), as we have seen in chapter 2, argued that Darwin was a species taxa realist, going as far as to characterize him as a precursor of the species-as-individuals view. When it came to the species category, however, Ghiselin characterized Darwin as a nominalist. Ghiselin's justification for this conclusion hinges on his claim that Darwin had an Aristotelian concept of category, a concept characterized by strict essentialism.[1] Since varieties, in Darwin's view, gradually evolve into species if they do not first go extinct, then contrary to the creationists there cannot be an essential distinction between species and varieties. So this is why the term "species" cannot on Darwin's view be defined. As Ghiselin (1969) put it, Darwin "insisted on Aristotelian definition as a criterion of reality or naturalness. To Darwin, as to many other taxonomists, an inability to give rigorous definitions for the names of taxonomic groups led to a belief that somehow such assemblages were artificial" (82). Ghiselin claims further that Darwin "maintained that there are no 'essential' differences between species and varieties, and that both terms designate the same basic kind of entity." Thus, says Ghiselin, "Darwin was denying the reality, not of taxa, but of categories" (93).

There are a number of problems with this view. First, as we have seen, Darwin did *not* think that species and varieties are basically the same kind of entity. Although they have a lot in common in Darwin's view, from the perspective of the horizontal dimension (and that as we have seen is the dimension for Darwin in which species have their primary reality) there is a fundamental distinction, most notably natural selection and the fixity of adaptations.

Second, there is no reason to believe that Darwin subscribed to Aristotelian essentialism, with the corresponding concept of definition (genos and differentiae), as "a criterion of reality or naturalness." For a start, Darwin never read Aristotle (cf. Burkhardt and Smith 1988, 498 n. 9). If he subscribed to Aristotelian essentialism for the reality of categories, then, his knowledge of Aristotle must have been secondhand. This knowledge was, of course, part of the worldview that Darwin inherited. Roughly a century before Darwin's *Beagle* voyage, Linnaeus wedded the Aristotelian paradigm to Christian creationism. This would be the received view for over the next hundred years (Stamos 2005). It was a view, as we have seen in chapter 1, that involved essentialism for membership in categories and bridgeless gaps between those categories, conceptualized as boxes, with accidental (nonessential) characters being irrelevant.

Although Darwin had indeed read Linnaeus, and of course the latter's worldview was (roughly speaking) the established view among Darwin's scientific contemporaries, at least for species, there is nevertheless nowhere in Darwin's writings where he explicitly subscribes to Aristotelian or Linnean criteria for the reality of categories. Granted, he makes statements which could be interpreted in that way. For example, in *Natural Selection* he says "it is no wonder that there should be difficulty in defining the difference between a species & a variety;—there being no essential, only an arbitrary difference" (Stauffer 1975, 98). Moreover, as we have seen earlier, near the end of the *Origin* he says "we shall at least be freed from the vain search for the undiscovered and undiscoverable essence of the term species"

(485). But these statements are no more to be taken at face value than his nominalistic claims about species taxa. The fact is that Darwin, as we have seen, in spite of his evolutionary gradualism, thought that species taxa are real. And it is especially important here to pay attention to his language. As we saw in chapter 3, Darwin in the *Origin* expressed his belief that "species come to be tolerably *well-defined* objects, and do not at any one period present an inextricable chaos of varying and intermediate links" (177, italics mine). Moreover near the end of the *Origin* he concluded that "Systematists will have only to decide (not that this will be easy) whether any form be sufficiently constant and distinct from other forms, to be capable of *definition*; and if *definable*, whether the differences be sufficiently important to deserve a specific name" (484, italics mine). And again, as we have seen in *Descent* (1871 I), he says "Every naturalist who has had the misfortune to undertake the description of a group of highly varying organisms, has encountered cases . . . precisely like that of man; and if of a cautious disposition, he will end up by uniting all the forms which graduate into each other as a single species; for he will say to himself that he has no right to give names to objects which he cannot *define*" (226–227, italics mine).

In all of this there is no evidence that Darwin was using the term "definition" in an Aristotelian sense. Quite the contrary, all he meant was *description*. Hence his use in the *Origin* of the phrase "good and distinct species" (47, 61, 259). All he was referring to were species that are distinct enough to be described, as in a species monograph. And Darwin, of course, had plenty of experience doing that, given his eight years of work on barnacles. Moreover there was nothing idiosyncratic in his use of the term "definition." For example, Hugh Strickland, in his first attempt at rules for zoological nomenclature (Strickland 1837), wrote in explanation of his Rule 7 ("A name may be expunged which has never been clearly defined") that "Unless a group is defined by description or figures when the name is given, it cannot be recognised by others; and the signification of the name is consequently lost. . . . Many collectors of shells and fossils are in the habit of labelling those species which they do not find described, with names of their own invention; but, unless they publish descriptions of these new species, they cannot expect these names to stand" (174).

Adding to all of this the evidence from chapters 2 and 3 that Darwin was a species taxa realist, it becomes evident that Darwin's concept of definition for individual taxa allowed for the reality of taxa even though his theory of gradual evolution precluded essentialism for those taxa. We can only conclude, then, that when it came to taxa Darwin had a realist though non-Aristotelian concept of definition. It seems odd, then, that Darwin would not extend this non-Aristotelian concept of definition to categories as well. But there is no reason to conclude that he did not. In fact the principle of charity dictates that we ascribe to Darwin (even if only at an intuitive level) a non-essentialistic concept of definition for the objective reality of categories, consistent with his non-essentialistic concept of definition for the objective reality of taxa. And if we do this, we place Darwin on perfectly logical ground. Strictly essentialistic categories, categories that imply bridgeless gaps among other things, are not the only kind of realist categories. Night grades into day and day into night, and not only are particular days and nights real but so are the two categories of night and day, yet neither category is strictly essentialistic.

By ascribing to Darwin a non-Aristotelian concept of definition and categories, we also place Darwin closer in a conceptual sense to the modern day. This is because it is common today among biologists who proffer or subscribe to a realist species concept to acknowledge that because of gradual evolution there are going to be "messy situations" in the application of their particular species concepts. A good example is Ernst Mayr. As we saw in chapter 2, Mayr argued that the species category, as defined by the biological species concept, is extra-mentally real. And yet he repeatedly acknowledged the existence in the biological world of what he called "messy situations." For example, Mayr (1982) states, in response to the often-repeated criticism that the biological species concept does not work well with plants (which we have seen in chapter 6), that "To be sure there are 'messy' situations . . . but I am far more impressed by the clear distinction of most 'kinds' of plants I encounter in nature than by the occasional messes" (285). Moreover Mayr later conducted a study on a local flora in the township of Concord, Massachusetts (Mayr 1992), in which he concluded that hybridization accounts for messy situations in 9.2% of the Concord flora (= 90.8% "good" species), which he thought might be reduced to 6.6% following further study (236). This analysis is in close accordance with McDade's (1995) survey of 104 recent botanical monographs, covering 1,790 species (mostly sexual), in which she concludes from her data that "something on the order of 15% of species are, at this horizontal time slice, involved in one or more biological processes that blur species boundaries or lead to infraspecific (and subsequently specific?) differentiation" (614). Moreover she concludes that "Variation was reported as sufficient to cause problems in delimiting about 7% of species treated" (613). Darwin was therefore justified, from a modern point of view, in stating his belief in the *Origin* that "species come to be tolerably well-defined objects, and do not at any one period present an inextricable chaos of varying and intermediate links." All things considered, his belief in the relatively low proportion of messy situations at any one horizontal level did not preclude species category realism on his part any more than it does for modern biologists with their various definitions of "species."

Another strategy theory is provided by Mayr (1982). Mayr takes Darwin's nominalistic statements on species, examined in chapter 1, pretty much at face value, so that for Darwin species are "purely arbitrary designations" (269). Moreover Mayr points the finger at botanists (the bane of Mayr's own species concept) for influencing Darwin toward species nominalism, especially William Herbert, whom he quotes as having written "there is no real or natural line of difference between species and permanent or discernible ['descendible' in the original] variety" (268). But the main thrust of Mayr's view is that Darwin had a "strong, even though perhaps unconscious, motivation . . . to demonstrate that species lack the constancy and distinctiveness claimed for them by the creationists" (269). On the prevailing creationist view, species must have distinct boundaries, even if difficult to determine by humans, for God in his perfection would surely not create a world with, or with the potential for, a confusion of forms. Indeed as Darwin put it in the *Origin*, "The view generally entertained by naturalists is that species, when intercrossed, have been specially endowed with the quality of sterility, in order to prevent the confusion of all organic forms" (245). Of course such boundaries would make no sense from the viewpoint of gradual evolution by natural selection. So Darwin, according to Mayr, had to deny the boundaries in order to make a case for his view of evolution. "Hence," he says, "it was good strategy to deny the distinctness of species" (269). This is why Mayr thinks Darwin focused on degree of difference rather than on reproductive isolation. By doing so he could make a more convincing case for species being "purely arbitrary designations" and thus for evolution.

There are a number of problems with Mayr's claims. First, there is a tension between his claim that Darwin was a species nominalist and his further claim that for Darwin "species continue to evolve" (269). One wonders how species can do this if they are "purely arbitrary designations."

Second, as we have seen in detail in chapter 1, in spite of William Herbert virtually all botanists in Darwin's time (and not just botanists) were species realists.

Finally, there is equally no reason to accept Mayr's claim that Darwin focused on degree of difference rather than on reproductive isolation as a matter of strategy. As we have seen in Chapter 6, Darwin rejected reproductive isolation, principally in the form of sterility, because he did not think that such a character could be formed by natural selection. Accordingly, species defined in terms of reproductive isolation would not on his view be part of a *vera causa* system of classification.

The most influential strategy theory has been provided by John Beatty (1985),[2] which is an improved version of his previous work (Beatty 1982). Because of its influence, it is important to examine it in some detail. To begin, Beatty (1985) acknowledges "Frank Sulloway's suggestion that Darwin's choice of species concept was guided by 'tactical' considerations. Among those tactical considerations was the decision to employ his fellow naturalists' species concept, in order to speak to them 'in their own language'" (265). According to Sulloway (1979), most of Darwin's fellow naturalists were "morphologists" and "typologists." Darwin therefore adopted a "morphological species definition" (38) as part of his tactic. In stressing the arbitrary distinction between species and varieties, "Darwin spoke to them in their own language and tailored his argument so as to exploit the inevitable sense of frustration that virtually all naturalists had at one time or another experienced in their taxonomic work" (37).

Beatty also acknowledges his debt to Ghiselin's (1969) interpretation of Darwin as a species taxa though not a species category realist. Beatty explicitly accepts this aspect of Ghiselin's thesis. In his own words, "Darwin's references to the arbitrariness and unreality of species pertained only to the species *category*, not to species *taxa*. . . . he did not deny the reality of the various species taxa like the cabbage and the radish" (277). As for species taxa, he also agrees with Ghiselin that such taxa were for Darwin "chunks with[in] the genealogical nexus of life" (278). Thus, for Beatty, Darwin's theory was not rendered domainless; in spite of its apparent nominalism, it still had referents, the individual species taxa. In going so far with Ghiselin, Beatty did not recognize, or at least acknowledge, what I claimed in chapter 2, namely, that such a view ascribes to Darwin a fundamental contradiction: if no species category, then "species taxa" becomes a contradiction in terms. Although Beatty does not recognize a contradiction, he does nevertheless recognize a fundamental tension. Near the end of his paper he says

> The suggestion that natural history could really get by without definitions of the categories of classification—especially a definition of "species"—is admittedly hard to swallow. Of course, it should be acknowledged that natural history was only temporarily without a definition of "species." The non-evolutionary definitions rejected by Darwin have since been replaced. Definitions such as Ernst Mayr's "biological species concept" and George Gaylord Simpson's "evolutionary species concept" are already so well entrenched as to be considered traditional. [280]

So far there is nothing really new in Beatty's theory. Where he departs from Ghiselin is in the referential use of the term "species" that he ascribes to Darwin and in the motive behind that use. We have already examined this aspect of Beatty's theory in chapter 5, but it is useful to look at it again. In Beatty's view, Darwin knew that his theory of evolution by natural selection was going to be a hard sell. According to Beatty, in writing the *Origin* Darwin was not trying to sell his theory to the general public. No Robert Chambers was he. Instead, he "addressed his theory" to the "community of 'naturalists,' of which Darwin considered himself a member" (266). The problem was that the species concept that these naturalists used was so theory-laden with non-evolutionary connotations that he could not possibly use their definition. But to replace it with an evolutionary species concept would have only made it more difficult, from a psychological point of view, to get his theory across to these naturalists. So he took the more subtle route of denying any reality to the species category, all the while throughout the *Origin* employing the word "species" in such a way so that it "agreed with his fellow naturalists' actual referential uses of the term" (269). As Beatty more succinctly put it,

> Darwin indeed perceived the difficulty posed by definitional language rules that undermined the theory he wished to communicate. He tried to get around this difficulty by distinguishing between what his fellow naturalists *called* "species" and the non-evolutionary beliefs in terms of which they *defined* "species." Regardless of their definitions, he argued, what they *called* "species" evolved. His species concept was therefore interestingly minimal: species were, for Darwin, just what expert naturalists *called* "species." By trying to talk about the same things that his contemporaries were talking about, he hoped that his language would conform satisfactorily enough for him to communicate his position to them. [266][3]

But Beatty goes even further than this. Accepting the quotations examined in chapter 1, where Darwin states that the term "species" is undefinable, Beatty claims that Darwin did not merely feign species category nominalism but actually meant it. In rejecting the species concepts of his fellow naturalists and employing only the referential use of those naturalists, says Beatty, it "allowed Darwin to communicate the position that the term 'species' was undefinable. In other words, Darwin not only rejected non-evolutionary definitions of 'species,' he also rejected the idea that the term could be defined at all" (274). Since Beatty clearly thinks that the term "species" can indeed be defined from an evolutionary perspective, given his reference to Mayr and Simpson above, it follows that he thinks that Darwin was just plain wrong.

I have argued in previous chapters that Darwin did indeed have an evolutionary species concept, albeit implicit. This saves Darwin from being wrong, not necessarily in his particular species concept, but in thinking, as Beatty suggests he thought, that gradual evolution precludes a realist species concept. At any rate, given that Beatty with his strategy theory goes against the (older) received view so far as to accept that Darwin really was a species taxa realist, one has to wonder why he does not extend his strategy theory further and argue that Darwin only feigned species category nominalism, that he was a species category realist after all. No doubt part of the reason is that Beatty, like everyone else, has failed to discern a species concept in Darwin's mature writings. But there is a further reason, much more overt, which, however, I must save for later in this chapter when I examine Hooker's conversion.

There is another aspect of Beatty's theory that I must save for later. His theory is clearly part of the larger incommensurability thesis made famous by Thomas Kuhn (1970), according to which adherents of old and new theories during a scientific revolution can barely if at all communicate their ideas between each other because of the theory-laden nature of their respective terminologies. I shall explore this issue, including Beatty's solution with regard to Darwin, in the next chapter. For the present, I want to go through the many reasons why I think Beatty's strategy theory should be rejected. This is necessary in order to clear the path for my own strategy theory.

Beginning with problems with Beatty's strategy theory, then, it is questionable whether Darwin wrote the *Origin* exclusively or even primarily for the community of his fellow naturalists. There is evidence, both within and outside the *Origin*, that he intended from the beginning a much wider audience. For a start, Darwin defines the term "endemic" in "endemic species" (390), something his fellow naturalists would surely not need done for them. Indeed the book is remarkably reader friendly all round. Moreover the *Origin* is filled with euphemisms designed, it would seem, for the greater sensibilities of the general public. For example, to play down the shivering coldness and utter horror of natural selection in the wild, Darwin wrote "When we reflect on this struggle, we may console ourselves with the full belief, that the war of nature is not incessant, that no fear is felt, that death is generally prompt, and that the vigorous, the healthy, and the happy survive and multiply" (79). Elsewhere he wrote of "the beautiful and harmonious diversity of nature" (169). Comforting words indeed, but Darwin's fellow naturalists would have known better. Even many amateur naturalists would have known better. Darwin himself, of course, knew better. In a letter to Hooker (July 13, 1856), for example, he wrote "What a book a Devil's chaplain might write on the clumsy, wasteful, blundering low & horridly cruel works of nature!" (Burkhardt and Smith 1990, 178). Similarly in a letter to Asa Gray (May 22, 1860) he wrote "There seems to me too much misery in the world. I cannot persuade myself that a beneficent & omnipotent God would have designedly created the Ichneumonidæ with the express intention of their feeding within the living bodies of caterpillars, or that a cat should play with mice" (Burkhardt *et al.* 1993, 224). There is further evidence of an intention for a wider audience that is even more direct. Interesting is Darwin's correspondence with his publisher, John Murray. Although Darwin admitted in one of his letters to Murray (April 2, 1859) that "My volume cannot be mere light reading, & some parts must be dry & some rather abstruse," he immediately added that "*yet as far I can judge perhaps very falsely*, it will be interesting to all (& they are many) who care for the curious problem of the origin of all animate forms" (Burkhardt and Smith 1991, 277; cf. 278, 440, and Burkhardt *et al.* 1993, 240). Given the high sales through many editions of Robert Chambers' anonymous *Vestiges of the Natural History of Creation*, beginning in 1844, Darwin could not have helped but know that there would indeed be "many" among the general public who would read his book.

A second and more important problem with Beatty's theory concerns his claim about Darwin's referential use of the term "species." I have argued extensively in chapters 4, 5, and 6 that Darwin did *not* simply follow the referential use of his fellow expert naturalists with regard to this term. In fact, most instructive are the cases where Darwin went against that referential use. Indeed it was those cases that largely helped form my reconstruction of Darwin's species concept.

Third, again in reference to Darwin's referential use of the term "species," it must be stressed that Darwin's fellow naturalists often did not provide him with a general consensus of referential use of species names. In many cases *they themselves could not agree*. Nowhere was this more evident to Darwin than in the case of barnacles. Characteristic of his frustration was what he wrote in a letter to J.S. Henslow (July 2, 1848):

> I am anxious to make out the distribution of the British species—And new species may turn up, for the group has been made out most superficially,—for instance under Balanus punctatus (which must be made a distinct genus) three or four varieties have been called distinct species; whereas one form, which has not been called even a variety, is not only a distinct genus, but a distinct sub-family.—Yesterday I found four or 5 named genera are all the closest species of one genus: this will give you a specimen of the utter confusion my poor dear Barnacles are in. [Burkhardt and Smith 1988, 156]

The situation was more widely reflected in the reality of species lumpers versus splitters. Even lumpers and splitters each among themselves often could not agree. As H.C. Watson put it in a letter to Darwin (March 24?, 1858), after tabulating the number of species variously allocated to six genera by eight botanists, "Thus, whether we compare together different Authors publishing at nearly the same dates, or the same Author publishing at different dates, much discrepancy appears in their ideas of species. . . . they [species] are far indeed from being fixed and certain in books" (Burkhardt and Smith 1991, 55–56). Far from providing Darwin with a referential consensus, then, Darwin's fellow naturalists often provided him with the opposite. Moreover they even lacked a consensus on the definition of the term. Indeed this is a theme that is repeated in the *Origin*. For example, Darwin says "No one definition has as yet satisfied all naturalists; yet every naturalist knows vaguely what he means when he speaks of a species" (44). Again, with regard to forms that naturalists could not agree on whether they are varieties or species, he says "to discuss whether they are rightly called species or varieties, before any definition of these terms has been generally accepted, is vainly to beat the air" (49). The theme would later be repeated in *Descent* (Darwin 1871 I), wherein, on the issue of whether the human races constitute one or several species, he says "it is a hopeless endeavour to decide this point on solid grounds, until some definition of the term 'species' is generally accepted; . . . We might as well attempt without any definition to decide whether a certain number of houses should be called a village, or town, or city" (228). Given, then, the already prevalent disagreements on the referential use of the term "species," and even on its definition, it would *not necessarily* have been a great impediment to the acceptance of Darwin's evolutionary views were he to redefine the concept.

A fourth problem with Beatty's theory, and the final one that I shall focus on, is his suggestion that Darwin not only learned of the problem of the theory-ladenness of species concepts from the writings of Watson (1845a, 1845b), but that Watson's writings "might also have suggested to Darwin a means of dealing with that difficulty" (271), which is the same means that Beatty attributes to Darwin. Much of Beatty's discussion on Watson cannot be contested. There is an error nevertheless, and the crux of it is in his use of the following quotation from Watson (1845a). Beatty writes that Watson "proposed to 'write of "species" as commonly understood by botanists, without attempting any rigorous definition of the term, which may hereafter be found to represent only a fiction of the human mind' (1845a, p. 142)" (273). In spite of the drift of the last part of this passage, it is important to emphasize that nowhere in Watson's writings, in

spite of his conversion to evolutionism apparently as a result of Chambers'
Vestiges, does he clearly and unequivocally advocate species category nominalism *or* species taxa nominalism. To the quotation above he prefixes it
with the words "For the present I must write of 'species' as . . . ." Moreover
he uses the word "may" when he writes of the word "species" representing
a fiction of the mind. He does not actually say that it *is* a fiction. In fact,
as we have already seen in chapter 1, although Watson came close at one
point to embracing species nominalism, he never went all the way, and
continued to think of real species in nature. He did indeed think of categories of higher taxa as indeed fictions of the mind (which was a common
view), but he did not think the same for the species category. Moreover
this was made clear to Darwin not only in Watson's writings but also in
Watson's correspondence to Darwin. There remains no reason, then, to
think that Darwin was following a strategy laid out earlier by Watson.
There was no such strategy, and what Darwin did was entirely different
from what Beatty suggests.

Parasitic on Beatty's theory is the theory of M.J.S. Hodge (1987).
According to Hodge, "only for some twelve months (late 1837 to late 1838)
did he [Darwin] articulate and uphold a definition of species. Before that
and afterwards he was consistently content not to attempt to contribute
one more to the growing number of definitional proposals in the literature" (233). During the period mentioned by Hodge, Darwin explicitly
endorsed, as we have seen in chapter 1, a species concept based on reproductive isolation. According to Hodge, this ceased once Darwin formu-
lated his theory of natural selection, which Hodge places in 1838 "in late
November and early December (not, contrary to legend, in September)"
(240).[4] Hodge explicitly follows Ghiselin in ascribing to Darwin, after this
period, species category nominalism combined with species taxa realism,
and he explicitly follows Beatty in ascribing to Darwin no change in the reference (denotation, extension) of the term "species" from his fellow naturalists. But Hodge adds to Beatty's theory a new twist. Beatty (1985) argues
that with regard to the non-evolutionary theory and corresponding definition of "species" of his fellow naturalists, Darwin had "at least two op-
tions": (i) he might respect the language rules of his fellow naturalists and
agree that the term "species" is to be used only for non-evolutionary entities but "object that there is nothing in the world that actually satisfies the
definition of the term," or (ii) he might respect not the language rules but
the *practices* of his fellow naturalists, the "*examples* of his fellow community
members' use of the term 'species'," but object to the use of the term for
only non-evolutionary entities, such that "the old theory about species, and

the theory-laden definition of the term, are substantial mischaracterizations of things that the community members have actually *called* 'species'" (268). According to Beatty, Darwin chose the second option because he thought it would be better for communicating his theory of evolution and for converting his fellow naturalists to his view. According to Hodge, Darwin did not even have a choice. As Hodge puts it, "Darwin was not setting out merely to find causes for those entities that were included in the reference of the term species. He was setting out to find causes for those properties that those entities had by virtue of which properties they counted, according to accepted criteria, as species and not as varieties. So Darwin's problem situation presupposed acceptance of the way the meaning or intension of the term *species*, as given by those criteria, had settled, determined, the reference of the term as it had done" (249).

Since Darwin did not, as I have shown, simply follow the referential use of his fellow naturalists with regard to species designations, his motive could not have been, contrary to what Hodge thinks, merely to determine the cause or causes of the specific characters as determined by his fellow naturalists. As we saw in chapters 4, 5, and 6, Darwin did not always follow species designations but often revised in accordance with his own evolutionary species concept. Moreover some of the characters that his fellow naturalists thought to be specific were accepted by Darwin if they had been fixed by natural selection. But other characters were rejected, such as sterility. Ghiselin (1969) points out that with Darwin "Classification ceased to be merely descriptive and became explanatory" (83). To my mind it became more, much more, to the point of actually *altering* some of the classifications of his contemporaries.

A further twist on Beatty's theory is provided by Gordon McOuat. McOuat (1996, 475, 2001, 3 n. 9) explicitly relies on Beatty's theory. Moreover McOuat makes much of what he (McOuat) calls, following Kitcher, the "cynical definition of species," which he says "was Darwin's *only* 'definition' of species" (1996, 515), the "only one clear definition of species in any of his work" (2001, 4 n. 10), explicitly given in *Natural Selection* and echoed in the *Origin*, viz., "In the following pages I mean by species, those collections of individuals, which have commonly been so designated by naturalists" (Stauffer 1975, 98). (Interestingly, when quoting this passage McOuat left out the word "commonly.") For McOuat, Darwin was not trying to bypass the theory-laden definition of "species" held by the majority of his fellow naturalists. For one thing, McOuat is skeptical that there was any consensus there at all. As he puts it (1996), "against the usual historical account, *it is unclear whether there really was a monolithic definition of species accepted by Darwin's fellow naturalists, or even the hope of one.* Moreover, science may not work like that anyway. Definitions are not at the core" (475). Instead of definitions, according to McOuat, what was important to Darwin's contemporary naturalists was "property rights," in particular the issue of whose nomenclature for taxa got to stick in the language community of the naturalists. This was especially important given the different systems of classification and nomenclature that were competing for acceptance at the time. Hence the importance of Strickland's (1837) initial rules for zoological nomenclature and later for the rules of a committee drawn together and reported by Strickland (1842). Of this committee, says McOuat (1996), "it was Darwin who convinced Strickland to co-opt the new members, from his own circle of contacts. Darwin soon became the main conduit between members, . . . Strickland was the organizational force behind the new commission, Darwin the fulcrum" (507). Of these rules, the law of priority would be "the one prime directive" (510). But more than that, says McOuat, in this publication of rules "the species category itself is not defined. This was not an 'essentialist' species concept, species with real 'definitions,' 'creationist' species, 'permanent' species, or whatever our histories have led us to believe. Rather, the solid core of conventional names marked something unsaid, yet something 'agreed' upon on trust: species are nodes in the communication of a network of 'competent' naturalists" (511). Thus for McOuat the stability of species came not from nature but from the need "to maintain the stability of naturalist discourse" (511). More specifically, "It was the restraint of language—the necessity of keeping names and their referents while all else, including definitions, remained 'up in the air'" (515). And it was this consideration, says McOuat, which explains Darwin's conservatism in accepting the species designations of his fellow naturalists. As he puts it, "Darwin, ever cautious, ever the conservative of revolutionaries, signed the document, knowing that soon he was to invoke the greatest change of 'meaning' while at the same time faithfully retaining the inherited names and stabilities of natural history and its objects" (514).

This is an interesting strategy theory indeed. Unfortunately it fails on at least two counts. First, in Strickland (1842) "the true system of nature" is assumed and the rules are meant to apply only "when naturalists are agreed as to the characters and limits of an individual group or species, [and when] they still disagree in the appellations by which they distinguish it" (106). Moreover there is a repeated appeal to follow the Linnean binomial system, going so far as to say "Two things are necessary before a zoological term can acquire any authority, viz. *definition* and *publication*. Definition properly implies a distinct exposition of essential characters, and in all cases we conceive this to be indispensable, . . . *publication*, nothing short of the insertion of the above particulars in a *printed book* can be held sufficient" (113–114). This points to what we have seen in Chapter 1, that in spite of the variety of species definitions competing at the time there was a common core, a core involving realism and what Wollaston called an "axiom" involving independent creation and non-transition. The different definitions involved, instead, different diagnostic criteria. Moreover I have argued in the previous chapter and elsewhere (Stamos 2005, 92–93) that there was a common core involving an underlying causal essence. J.D. Dana (1857), for example, expressed this view when he compared biological species to chemical elements, the main difference being that individuals of a biological species reproduce, such that "when individuals multiply from generation to generation, it is but a repetition of the primordial type-idea; and the true notion of the species is not in the resulting group, but in the idea or potential element which is at the basis of every individual of the group" (487). Ellegård (1958, 210) calls this the traditional view, and I would concur.

Secondly and equally important, McOuat makes the same mistake as the others, and that it is to assume that Darwin simply followed the species designations of his fellow naturalists. But as we have seen in previous chapters, this turns out not to be true. The square peg does not fit the round hole. Another theory is therefore needed.

Finally, there is my strategy theory published in Stamos (1996). Unlike Ghiselin and Beatty before me, I felt guided by the principle of charity to reject their view that Darwin was a species taxa though not a species category realist. This seemed to me to attribute to Darwin a fundamental confusion: "No species category, no *species* taxa" (133, italics added). To say there are species taxa but no species category is to make out of the phrase "species taxa" a contradiction in terms. It didn't seem fair to attribute to Darwin such a confusion. Nevertheless instead of trying to reconstruct Darwin's realist though implicit definition of the species category, I concluded that "there appears no good reason (contra Beatty) to believe that he [Darwin] would have denied the reality of the species category, although he may not have had a sufficiently clear idea in order to define it. But, of course, it need only be remembered that to define it would have been for Darwin to contradict and defeat what needed to have been a lifetime strategy" (144). "Needed to have been a lifetime strategy"? I still think so, though not for the same reasons. "May not have had a sufficiently clear idea in order to define it"? I now sincerely think otherwise. If a theorist inconsistently uses grouping and ranking criteria, then perhaps a good case can be made for that person not having a clear idea. But when a theorist, especially of Darwin's caliber, repeatedly and consistently uses the same grouping and ranking criteria, then a strong case can and should be made for that person having a clear idea. In the case of the species category, to say that Darwin did not believe that the species category is real is to say that he did not believe that there are objective and universally valid criteria for delimiting species (objective and universal in the sense of the four criteria discussed in chapter 2). Since, as I believe, I have in previous chapters reconstructed to a reasonable degree what those criteria were for Darwin, a strategy theory, in particular one that fits the facts, is needed in order to complete the picture.

But if all of the foregoing strategy theories are unacceptable, then what is left? If Darwin was not only a species taxa realist but also a species category realist, with an actual though implicit definition of the term "species" fully in accordance with his theory of gradual evolution by natural selection, why then would he *explicitly* maintain the undefinability, and by implication the unreality, of the species category? This cries out for an explanation. Moreover, it cries out for an explanation that fits as much of the evidence as possible and that extends to Darwin the principle of charity to the greatest possible degree.

Such an explanation cannot possibly hope to ignore the fact that the received view in Darwin's time, prior of course to the *Origin*, was that species were created by a First Cause (God), either once or in a series of successive creations, and that varieties were created by secondary causes (natural laws). On this popular view the reality of species depended on this essential distinction. If not only varieties but also species were created by secondary causes, so that varieties are incipient species, this entailed not only that species are indefinitely mutable but that species are no more real than varieties, which in turn impugns the reality of both species taxa and the species category. And indeed we have seen this in chapter 1. Lyell (1832), Whewell (1837), Watson (1845b), and Wollaston (1860) are prominent examples of the view—what may be taken to have been during that time the prevalent view—that either species are fixed, in which case they are real, or they are not fixed, in which case they are not real.

In defining both species taxa and the species category nominalistically, then, it seems to me—building on my theory first presented in Stamos (1996, 133–137)—that Darwin was not so much trying to communicate his theory of evolution as he was applying to his opponents a consistency argument:

P1: You already employ common descent in grouping males and females and larvae and adults and monsters and varieties as conspecific, in spite of often great dissimilarity.[5]

P2: You believe that species are real because they are fixed by divine fiat, but that varieties are not real because of their variability from secondary causes, so that what are called "varieties" are merely arbitrary groupings made for the sake of convenience.[6]

P3: New evidence and arguments strongly suggest, contrary to the received view, that what are often called "varieties" are incipient species (indeed this is the main theme of the *Origin*).[7]

C1: *You therefore must conclude that, just like varieties, species are merely arbitrary groupings made for the sake of convenience, that they are not real either.*

Given that almost all of Darwin's contemporaries were creationists of some sort (cf. chapter 1), this would have been a clever argument against them. It is an example of the *reductio ad absurdum* form of argument, not in the technical sense that one finds in books on formal logic, but in the informal sense. Whether Darwin knew of the name of this type of argument, I do not know, but certainly it is an intuitive form of argument that has been around for thousands of years. As Gorovitz *et al.* (1979) characterize this informal technique, "we show a position to be unacceptable by showing that it leads to—or can be reduced to—something absurd or clearly unacceptable for some other reason" (141). Having accomplished that, the technique forces the opponent to reject one of the original premises. In the above example, the conclusion that species are not real follows logically from the premises. And if the premises are true, then the argument is sound. But the conclusion that species are not real would not be acceptable to most of Darwin's contemporaries. Better to have species that evolve than no species at all. The paradigm shift would be difficult, but it would be better than to accept the conclusion that species are not real. The latter involved a taxonomic chaos that few could stomach. Lyell, for example, stated in a letter to Hooker (July 25, 1856) that as long as those who multiply species "feared that a species might turn out to be a separate and independent creation, they might feel checked; but once abandon this article of faith, and every man becomes his own infallible Pope" (Burkhardt and Smith 1990, 194 n. 6).[8] Moreover there was the religious difficulty. How could species be not real if God exists? Does God think not of species? Does not the Bible itself, in Genesis, speak of God creating different *kinds* of plants and animals? The only way out is to keep God and accept evolution, with species being in some different way real and their evolution being directed by the will of God. One of the above premises, then, had to go. And the obvious candidate for rejection would be P2. P1 was followed by all parties, so it could not go. P3 could go if one accepted that Darwin had not made a strong case for the fact of evolution. Since Darwin had thought he made a strong case (and this was indeed borne out over the decade following 1859), the only premise remaining, and the one that must be rejected and replaced, was P2.

If this was indeed Darwin's strategy, why then, one must wonder, did he not come out with his species concept once the *Origin* had done its job and converted most of the educated world in the West from a static concept of the biological world to an evolutionary one? The answer, I suggest, if my line of reasoning is correct thus far, is because of the central element of natural selection in his species concept, reconstructed in chapter 5. Although Darwin's *Origin* had converted most biologists in Great Britain (many of whom were Anglican clergymen), many other scientists, and much of the lay public to evolutionism in the space of a decade (Ellegård 1958; Hull *et al.* 1978), it never converted that world to natural selection as the main mechanism of species evolution. In fact, Darwin's theory of natural selection remained largely unaccepted among biologists (never mind the general public) until well after his death, beginning in the 1930s (Ruse 1979, 228–233). Darwin, then, may have perceived it as hazardous to proffer his own species concept, given that it was based on natural selection, until natural selection itself was generally accepted. Since natural selection was always far from being generally accepted during the remainder of his lifetime, he never bothered to make explicit his mature species concept. Moreover, given the already contentious nature of the definition of "species" from within the creationist framework, Darwin may also have thought (in a prescient way) that even if evolution by natural selection is eventually accepted by most biologists, they still will not be able to agree on a definition of "species." This is all, of course, pure speculation, eminently reasonable though lacking positive evidence, and I won't try to take it any further.

A further problem raised by my theory on Darwin's strategy concerns the concept of variety, what Darwin's fellow naturalists thought about the nature of varieties, whether they're real or arbitrary, and what Darwin himself thought. Clearly much of my theory hinges on this issue, so much so that I venture here to affirm that if my analysis in the previous chapter ultimately fails, then so too does my strategy theory developed in the present chapter. For the sake of presenting the theory, however, I shall assume that my analysis in the previous chapter is basically sound. Onward, then, with the theory.

Although there is an obvious problem with the record, since much of Darwin's correspondence is missing,[9] and we obviously do not have recordings of his conversations, a particular trend is nevertheless sufficiently perceivable to the attentive eye while moving up the strata of Darwin's correspondence. In case after case, it seems that Darwin employed a stance of species nominalism when dealing with an outstanding naturalist whom he thought he had a chance of converting to his views, which he then replaced with a stance of species realism once the conversion had become to his mind sufficiently complete. It is also significant that with each of these naturalists he never so much as even hinted at species nominalism prior to his confession of evolutionism.

A prime example is Asa Gray, Darwin's main exponent of his views in America. Prior to his conversion by Darwin, Gray had been for a long time a creationist and essentialist. According to Dupree (1959), Gray "took his main arguments" from Augustin-Pyramus De Candolle's *Théorie Élémentaire de la Botanique*, first published in 1813, in which a species is defined as "the collection of all the individuals who resemble one another more than they resemble others; who are able, by reciprocal fecundation, to produce fertile individuals; and who reproduce by generation, such kind as one may by analogy suppose that all came down originally from one single individual" (54). The species concept defined here, although it allows for variation and varieties, is clearly essentialist and precludes the evolution of one species from another (54). Gray first read De Candolle's *Théorie* in 1834 (40) and defended his species concept in his (Gray's) *Elements of Botany*, published in 1836. This species concept Gray retained and defended for many years, until he was converted to evolutionism by Darwin (217).

In a letter to Gray (July 20, 1857), Darwin first confessed to Gray his evolutionism, providing a very brief précis of his theory. In this letter Darwin did not touch on the subject of whether species should now be thought of as real or not. The most he says is "I have come to the heterodox conclusion that there are no such things as independently created species—that species are only strongly defined varieties" (Burkhardt and Smith 1990, 432). A little later in another letter to Gray (September 5, 1857) Darwin provided a more detailed précis of his theory, which was later read to the Linnean Society on July 1, 1858, along with Wallace's "On the Tendency of Varieties to depart indefinitely from the Original Type" (Barrett 1977 II, 3–19), which together were published in the Society's *Journal* in early 1859. But Gray was still by no means convinced. A little less than three months later than the previous letter Darwin would write to Gray (November 29, 1857) the following:

> When I was at systematic work, I know I longed to have no other diffi-
> culty (great enough) than deciding whether the form was distinct enough
> to deserve a name, & not to be haunted with undefined and unanswer-
> able question whether it was a true species. What a jump it is from a well
> marked variety, produced by natural cause, to a species produced by the
> separate act of the Hand of God. . . . By the way I met the other day
> Phillips, the Palæontologist, & he asked me "how do you define a
> species?"—I answered "I cannot" Whereupon he said "at last I have found
> out the only true definition,—'any form which has ever had a specific
> name'"! [Burkhardt and Smith 1990, 492–493]

In this letter Darwin clearly takes a stance of species nominalism. In quoting, apparently favorably, the paleontologist John Phillips, one might think
that Darwin was only advocating species nominalism over time, only vertical species nominalism. But what he wrote immediately prior to this passage suggests otherwise. Darwin clearly states that he thinks the term
"species" is not only "undefined" but that its definition is an "unanswerable
question." Combined with what he says about the difficulty in deciding
whether a particular form is a species or a variety, the gist of this letter is
nominalism for both species taxa and the species category.

Gray remained unconvinced. Interestingly in a letter written over a
month after the publication of the *Origin*, Darwin seems to repeat his
stance of species nominalism. In a letter to Gray (December 21, 1859) he
expressed his view that intelligent readers of his book who are not naturalists "will drag after them those naturalists, who have too firmly fixed in
their heads that a species is an entity" (Burkhardt and Smith 1991, 440).

According to Dupree (1959, 249–269) it was not until Gray had read
Darwin's *Origin*, primed for it mostly by Hooker's correspondence throughout 1859 and by his own earlier and recent work on the highly similar flora
of Japan and eastern North America, that Gray became in the main convinced. I am not quite so sure, however. At any rate, what matters for my
strategy theory is Darwin's *perception* of Gray's conversion. Gray had finished reading Darwin's *Origin* on January 1, 1860. He was duly impressed,
writing to Hooker on January 5 that "It is done in a *masterly manner*"
(Burkhardt *et al.* 1993, 16). However, in a letter to Darwin (January 23),
with regard to his first review of Darwin's book, he wrote that it is better for
Darwin's theory that he take a neutral position rather than announce himself as a "convert," to which he immediately added "nor could I say the latter with truth" (47). Accordingly, Darwin did not include Gray in his list of
converts. In a letter to Hooker (March 3) he listed Gray as a convert only
"to some extent" (116). However, Gray's extensive efforts on Darwin's

behalf (including three favorable reviews and public debates with Agassiz) were not lost on Darwin. In a letter to Lyell (July 5) he wrote "Huxley, Hooker & J. Lubbock (as I am pleased to hear) seem to have stuck up for modification of Species like Trojans. Asa Gray, as I hear today, also goes on fighting well" (281). Indeed in a letter to J.D. Dana (July 30) Darwin stated that "No one person understands my views & has defended them so well as A. Gray" (303). Similarly in a letter to Lyell, written on the same day, Darwin wrote "No one I think understands whole case better than Asa Gray, & he has been fighting nobly" (306). In the letter to Dana, Darwin immediately added the disclaimer "though he does not by any means go all the way with me." Gray's reservations were mainly theological. He could not accept that variations were random with respect to the environment. Instead he thought that favorable variations must be directed by God. Nevertheless Gray was enough of a convert, accepting the probability of both species evolution and natural selection as the main mechanism, and Darwin clearly thought highly of his efforts. Accordingly Darwin changed his language with respect to species reality in his correspondence with Gray. As we have seen in chapter 1, in reply to Agassiz's (1860b) quip "If species do not exist at all, as the supporters of the transmutation theory maintain, how can they vary?" (143), Darwin wrote to Gray (August 11, 1860) "How absurd that logical quibble;—'if species do not exist how can they vary?' As if anyone doubted their temporary existence" (317). In chapter 3 I attempted to elucidate Darwin's meaning of the word "temporary." At this point, however, we need not be concerned with that matter. Existence, however temporary, is still real existence, and Gray could hardly have exacted any other meaning from Darwin's letter (cf. chapter 1 note 8). Darwin's language here to his new convert, unlike his earlier letters written when Gray was still relatively orthodox, is clearly the language of species realism.

An equally striking though more complicated example is Darwin's correspondence with J.D. Hooker, not only for many years Darwin's closest friend but also his main correspondent. Darwin first confessed his evolutionism to Hooker in one of his letters (January 11, 1844), in which he stated "I am almost convinced (quite contrary to opinion I started with) that species are not (it is like confessing a murder) immutable" (Burkhardt and Smith 1987, 2). Roughly three years later, as indicated in one of his letters to Hooker (February 8, 1847), Hooker read Darwin's *Essay of 1844* (Darwin 1909). However, it is by no means easy to determine when Hooker became convinced. We know that by mid-1850 he was still not a convert. In a letter to Darwin (April 6 and 7, 1850) he wrote "I remember once dreaming that you were too prone to theoretical considerations about species & unaware of certain difficulties in your way, which I thought a more intimate acquaintance with species *practically* might clear up. Hence I rejoiced at your taking up a difficult genus [barnacles] & in a manner the best calculated to throw light on specific characters their value &c. Since then your own theories, have possessed me, without however converting me" (Burkhardt and Smith 1988, 328). As with Darwin's *Natura non facit saltum* claim for species, Hooker's conversion appears to have been exceedingly gradual. Peter Stevens (1997, 346) suggests that Hooker became a convert early on, already by the time that he was traveling in India (Hooker set sail for India on November 11, 1847, botanized for almost 3½ years there, and returned home on March 25, 1851). Desmond and Moore (1992, 416, 713 n. 28) suggest that Hooker's conversion was done by July 1855 (cf. Hooker's letter to Darwin, July 8, 1855, Burkhardt and Smith 1989, 372). I believe that Hooker's conversion was not settled until three years later. At any rate, what really matters for my strategy theory is Darwin's *perception* of Hooker's conversion. To begin, in a letter to Hooker (December 24, 1856) that we have seen previously, Darwin wrote

> I have just been comparing definitions of species, . . . It is really laughable to see what different ideas are prominent in various naturalists minds, when they speak of "species" in some resemblance is everything & descent of little weight—in some resemblance seems to go for nothing & Creation the reigning idea—in some descent the key—in some sterility an unfailing test, with others not worth a farthing. It all comes, I believe, from trying to define the undefinable. [Burkhardt and Smith 1990, 309]

Here we find Darwin clearly advocating species nominalism, at least for the species category. As we have seen in the previous chapter as well as in chapter 1, an almost identical passage was written at nearly the same time for the manuscript *Natural Selection*, perhaps even on the same day. When we take a closer look at the species concepts that Darwin is referring to in both of these passages, however, the duplicitous nature of both of these passages becomes quite apparent when viewed in the light of the preceding chapters. What is interesting is what is missing. In *Natural Selection* (Stauffer 1975, 95–97) Darwin discusses a number of species concepts. He discusses at length the species concept of Alphonse de Candolle, who considered resemblance more important than descent. Darwin rejects this using the case of primroses and cowslips. He then discusses the species concept of Louis Agassiz, who allowed for perfectly similar forms to be different species if found in widely separate localities. It is obvious how Darwin will reject this, when in his manuscript he gets to the topic of dispersion.

Finally, he discusses the species concept of Joseph Gottlieb Kölreuter, who took sterility between forms as the sole test. Darwin will reject this when he gets to the topic of sterility. What is missing in all of this is the kind of species concept closest to Darwin's own, as I reconstructed from his usage (e.g., dogs, cattle, primroses and cowslips, pigeons, humans). Specifically, what is missing is the kind of species concept advocated by Edward Blyth on domestic cattle and added to by Darwin in terms of descent. Indeed, it allows us to say how Darwin would have rated each of the criteria discussed in the passage quoted above, as I have done in the previous chapter.

At any rate, because this nominalistic passage is in a letter rather than a published work, Beatty (1985, 274–275) apparently took this passage in league with the others like it, as revealing Darwin's true beliefs, so that he did not feel a need to attempt a strategy theory for Darwin's apparent species category nominalism. I strongly suggest, however, that this passage from Darwin's letter to Hooker should be taken no more seriously than any comparable passage in his published works. For it seems to me that this letter may have been occasioned by the apparent fact that Hooker was still orthodox with respect to species. Only a month and a half prior to this letter, after having read Darwin's manuscript chapter on geological distribution (chapter XI in *Natural Selection*), Hooker wrote to Darwin (November 9, 1856) "Your case is a most strong one & gives me a much higher idea of *change* than I had previously entertained; &, though, as you know, never very stubborn about unalterability of specific type, I never felt so shaky about species before" (Burkhardt and Smith 1990, 259).

Although shaken, Hooker was still nevertheless not converted. At any rate, what matters here is Darwin's *perception*. Interesting is what Darwin wrote about Hooker in chapter IV of *Natural Selection*: "Dr. Hooker objects to my whole manner of treating the present subject because varieties are so ill defined; had he added that species were likewise ill defined, I should have entirely agreed with him; for my belief is that both are liable to this imputation" (Stauffer 1975, 159–160). Here Darwin clearly refers specifically to species taxa. It is obvious that he thinks that Hooker thought of species taxa as being real. But instead of agreeing (which we know he would have), he impugns their reality, taking a stance of species taxa nominalism. Darwin's reference to Hooker is either to a verbal communication or to a letter now lost.[10] But the significance is not lost. The implication is that Darwin felt that Hooker still believed to some degree in the definability of the term "species," in other words, the reality of the species category in some sense traditionally conceived. According to Stauffer (1975, 94–95), the chapter section from which this passage is taken was begun on March

10, 1858, and completed on May 14 of the same year. This means that it was completed roughly 16½ months *after* Darwin's letter to Hooker above (December 24, 1856), in which he claimed that the term "species" is undefinable. To this we may add Darwin's later reflection written in his autobiography (1876a):

> It has sometimes been said that the success of the *Origin* proved "that the subject was in the air," or "that men's minds were prepared for it." I do not think that this is strictly true, for I occasionally sounded not a few naturalists, and never happened to come across a single one who seemed to doubt about the permanence of species. Even Lyell and Hooker, though they would listen with interest to me, never seemed to agree. [123–124]

As his closest friend and confidant, Darwin was certainly in a position to be privy to Hooker's thoughts and feelings on the issue. Thus, although the above fits my strategy theory, the problem concerning Hooker's conversion needs to be resolved. Hooker would certainly not have withheld it from Darwin. Although Hooker may have teetered back and forth for many years, there must have come a point when he teetered back no more but rested on evolutionism. So was it as early as Stevens suggests, or a little later as Desmond and Moore suggest, or was it closer to the publication of Darwin's *Origin*? Hooker's first *public* statement of his conversion to evolutionism, and specifically that of the Darwin and Wallace variety, occurred in the introductory essay of his *Flora Tasmaniæ*, begun in mid-December 1857 (Burkhardt and Smith 1990, 508) and completed and sent to press in mid-March 1859 (Burkhardt and Smith 1991, 258), though its publication was delayed until mid-December 1859. According to Burkhardt and Smith (1991), "it was in the course of writing up this essay that Hooker became convinced of the value of Darwin's theory" (xx).[11] Interestingly, when Hooker began writing this essay he had no intention of expressing any conversion to evolutionism. Instead, as he wrote in a letter to Darwin (December 17–23, 1857), he intended "to confine it to a clear exposition of all the main features of the Flora of Australia & leave all conclusion drawing to others" (Burkhardt and Smith 1990, 508). In reply, Darwin wrote to Hooker (December 25) "I am very sorry to hear that you do not intend to give generalisations in your Tasmanian Introduction, but I do not believe you will be able to resist: what is in the spirit must come out" (516). And of course it did eventually come out, in December of 1859. But at what point during the writing of his introductory essay did Hooker become completely convinced?

The answer is apparently after reading Darwin's manuscript on large and small genera, which would become the second section of chapter IV in Darwin's *Natural Selection* (Stauffer 1975, 134–171), in which Darwin argued that large genera tend to have more varieties per species than small genera, the upshot being the strong support this generalization provides for his view that varieties are incipient species (163–164). Upon reading the manuscript, Hooker wrote to Darwin (July 13–15, 1858) "I was at Mr Smith's of Combe Hurst last Sunday . . . . I there went deep into your Mss on variable species in big & small genera . . . . After very full deliberation I cordially concur in your view & accept it with all its consequences" (Burkhardt and Smith 1991, 131). As if to confirm Hooker's conversion, Darwin would write a few months later to Hooker (October 20) "indeed I thought, until quite lately, that my M.S. had produced no effect on you & this has often staggered me" (174). Certain that Hooker's conversion was set, we then find Darwin write to his friends, starting in 1859, that "D^r. Hooker has become almost as heterodox as you [Wallace] or I" (January 25), that "I have had the great satisfaction of converting Hooker" (February 12), that Hooker "is a *full* convert" (April 5), "goes the whole length" (August 9), and "gives up species in grand style" (September 2), which is to say "has completely given up species as immutable creations" (November 11) (Burkhardt and Smith 1991, 241, 247, 279, 324, 330, 372).[12]

Having believed from Hooker's letter (July 13–15, 1858) that Hooker was now a complete convert, Darwin no longer used the language of species nominalism on him. For example, in a letter to Hooker (December 24, 1858) he states "what an important datum, it would be if one knew average comparative rate of specific change in Mammals & plants" (Burkhardt and Smith 1991, 222). This was in reply to Hooker's claim in his letter to Darwin (December 22) that "I also still regard plant types as older things than animal types" (219), in which he remained neutral on whether his fossil and living plant examples, "identical to all appearances," were specifically identical. Darwin would still use in his correspondence with Hooker during this post-conversion period the phrase "give up species," as for example when he wrote to Hooker (November 14, 1858) that it has long troubled him that he interfered with Hooker's own work by belaboring him with so many letters and questions and ideas, and added "My only comfort is, that without you were prepared to give up species" (199). But by this he only meant the immutability of species.

Another striking case is Darwin's correspondence with Charles Lyell. As we have seen in chapter 1, Lyell, as Darwin well knew, claimed in his *Principles of Geology* (Lyell 1832) that if species evolve, then they aren't real (cf. Coleman 1962). Moreover, Lyell retained this view for a considerable amount of time. For example, in a letter to Darwin (May 1–2, 1856), in reference to the weekend in late April 1856 during which Hooker, Huxley, and Wollaston stayed at Darwin's home, Lyell wrote "I hear that when you & Hooker & Huxley & Wollaston got together you made light of all species & grew more and more unorthodox" (Burkhardt and Smith 1990, 89). Moreover in a letter to Hooker, Lyell wrote that if naturalists believed that species are "artificial, or mere human inventions" rather than separate creations, then "once abandon this article of faith, and every man becomes his own infallible Pope" (194 n. 6). Interestingly, earlier in the month of April, the month that Hooker, Huxley, and Wollaston visited Darwin, Lyell himself had visited Darwin for a weekend. According to Burkhardt and Smith (1990, 91 n. 10), all the evidence, including Lyell's own notes, strongly suggests that it was at this time that Darwin revealed his evolutionism to Lyell, including his theory of natural selection. One would especially expect, then, given Lyell's ontological dichotomy on the species question, that Darwin would employ against Lyell the nominalism/realism strategy that I ascribe to him.

And this is indeed what one finds. In fact, one finds it well before Darwin confessed his evolutionism to Lyell (assuming Burkhardt and Smith are correct about when that was). In a letter to Lyell (July 30, 1837) Darwin wrote "Has your late work at shells startled you about the existence of species? I have been attending a *very* little to species of birds, & the passage of forms, do appear frightful—every thing is arbitrary; no two naturalists agree on any fundamental idea that I can see" (Burkhardt and Smith 1986, 32). This was written during Darwin's brief monadic theory of evolution, which makes Darwin's taunt even more interesting. Many years later, shortly after Lyell learned about Darwin's evolutionism (again, assuming Burkhardt and Smith are correct about when that was), Lyell strongly encouraged Darwin to publish his views and was instrumental in arranging the reading of the Darwin and Wallace papers to the Linnean Society on July 1, 1858. Nevertheless, he was far from being a convert. Of course the two had long been good friends, and Darwin respected Lyell's feedback to the extent that he had Lyell read drafts of chapters of the *Origin* before going to press. It is significant that in a letter to Lyell (September 2, 1859), in which Darwin briefly discussed the chapters he was having sent to Lyell, he wrote "I have read some of Hookers Introduction to Australian Flora, & he gives up species in grand style" (Burkhardt and Smith 1991, 330). Darwin had to have known that Lyell would have taken this as referring to species nominalism, a position that would be a logical consequence of

Lyell's dichotomy on species, that either species are fixed and real or not fixed and unreal.

Lyell's acceptance of the main theme of the *Origin*, however, is tricky to nail down. But again, what really only matters here is Darwin's *perception* of Lyell's conversion. In a letter to Wallace (November 13, 1859) Darwin wrote "No one has read it [the ms. of the *Origin*] except Lyell, with whom I have had much correspondence. Hooker thinks him a complete convert; but he does not seem so in his letters to me" (Burkhardt and Smith 1991, 375).[13] However, shortly after, the day before the official publication of the *Origin*, Darwin seems to have thought that Lyell had become a convert. In a letter to Lyell himself (November 23, 1859) he wrote:

> I rejoice profoundly that you intend admitting doctrine of modification in your new Edition [the next edition of Lyell's *Principles of Geology*]. Nothing, I am convinced, could be more important for its success. I honour you most sincerely:—to have maintained, in the position of a master, one side of a question for 30 years & then deliberately give it up, is a fact, to which I much doubt whether the records of science offer a parallel. [391–392]

Moreover throughout the month of December Darwin proudly proclaimed Lyell a convert to a number of his correspondents. In a letter to J.L.A. de Quatrefages de Bréau (December 5) he wrote "Sir C. Lyell, who has been our chief maintainer of the immutability of species, has become an entire convert" (Burkhardt and Smith 1991, 415). In a letter to Gray (December 24) he wrote "Lyell is a complete convert" (446). In a letter to his cousin W.D. Fox (December 25) he wrote "The Book has already made a few enthusiastic & first-rate converts, viz. Lyell, Hooker, Huxley, Carpenter &c." (449). And this continued on for quite some time. In a letter to Wallace (May 18, 1860) Darwin wrote "Lyell keeps as firm as a tower, & this autumn will publish on Geological History of Man, & will there declare his conversion, which now is universally known" (Burkhardt *et al.* 1993, 220). Accordingly, Darwin now uses with Lyell the language of species realism. For example, in a letter to Lyell (December 22, 1859) he wrote "I have heard from Sir W. Jardine: his criticisms are quite unimportant—some of the Galapagos so-called species ought to be called varieties, which I fully expected—Some of the sub-genera thought to be wholly endemic have been found on Continent (not that he gives his authority) but I do not make out that the species are the same" (Burkhardt and Smith 1991, 442).

Lyell, however, was in reality fence sitting, working ideas out in his species notebooks (Wilson 1970), and Darwin was misled. In fact, Lyell did not discuss Darwin's theory in print until *The Antiquity of Man* (Lyell 1863), which deeply disturbed Darwin because of Lyell's inconclusive discussion on the evolution of species, to the point of rejecting the evolution of man from another species. In fact, Darwin was shocked that Lyell did not come out for evolution, as he was convinced ("at times") that Lyell was in fact a complete convert (cf. Darwin's letters to Lyell, March 6 and 12–13, 1863, and to Gray, Hooker, and Huxley in the month before; Burkhardt *et al.* 1999, 207–209, 222–223, 166, 173–174, 181). In fact it was not until the tenth edition of his *Principles of Geology*, issued in two volumes in 1867–1868, that Lyell finally publicly accepted Darwin's view (Hull *et al.* 1978, 52)—although a few years earlier on the evening of November 30, 1864, in an after-dinner speech in honor of Darwin having been awarded the Copley Medal of the Royal Society, he gave what he called a "confession of faith as to the *Origin*" (Darwin 1892, 274).

What is interesting is that throughout the years 1860, 1861, 1862, and up until the publication of Lyell's *Antiquity*, Darwin sometimes suspected Lyell's fence sitting but never fully caught on. We can see this especially in the correspondence between Lyell and Darwin in 1860 (Burkhardt *et al.* 1993). In a letter to Hooker (March 3), he listed Lyell as a convert among geologists (116). In July he was looking forward to Lyell publishing on the subject of evolution (299), and then a little later he writes to Lyell of "our side" (306). By September, however, Lyell was talking about "Most species immutable & true to the death, as I maintained in the Principles that all were, but a few of the whole plastic & becoming the centers of new generic & ultimately higher groups as you maintain" (348), which Darwin took as a hint of "the creation 'of distinct successive types,'" such that "You cut my own throat, & your own throat" (356), to which Lyell replied "You need not be afraid of my starting any theory of successive creation of types" (366). Darwin then wrote to Gray (September 26) "I can perceive in my immense correspondence with Lyell, who objected to much at first, that he has, perhaps unconsciously to himself converted himself very much during last six months" (390). A day later, however, Lyell wrote to Darwin (September 27) saying he is "haunted with a kind of misgiving" (393) over Darwin's thesis of the multiple origin of domestic dogs. Darwin then replied (September 28) with a kind of chastisement, saying "I cannot see yet how multiple origin of dog can be properly brought as argument for multiple origin of man. Is not your feeling a remnant of that deeply-impressed one on all our minds, that a species is an entity,—something quite distinct from variety?" (397). In this passage there is a hint of the language of species nominalism, but it would not come out fully quite yet.[14] Two days later

Lyell in a letter to Darwin (September 30) reiterated his theme of immutable species, calling them "a thousand to one" compared to the mutable kind, which he wrote "accords with my old notion that species as a rule, or the majority of them, are immutable, & have always been so" (399). Consequently we find Darwin (October 8) hint to Lyell that he is still a creationist (421). And then in a further letter to Lyell (December 4) Darwin clearly returns to the language of species nominalism: "How far to lump & split species is indeed a hopeless problem.—It must in the end, I think, be determined by mere convenience" (512).

Turning to 1861 and 1862, Lyell had been working on the sixth edition of his *Manual of Elementary Geology* (which would not be published until 1865), and in a letter to Darwin (before August 21, 1861) he appended a note that he intended to add to the *Manual*, which was more conciliatory to Darwin's views (Burkhardt et al. 1994, 239 n. 2). Lyell's letter (not extant) was apparently in agreement with Gray's view that God guides favorable mutations in the evolution of species, and Darwin in his reply letter to Lyell (August 21) apparently took the letter and the note to mean that Lyell was "one of the wretches" (237), meaning a believer in evolution. Accordingly we find Darwin writing to Gray (September 17) "I have been lately corresponding with Lyell, who, I think, adopts your idea of the stream of variation having been led or designed" (267). In a letter to Lyell (August 22, 1862) Darwin expresses his hope that Lyell's *Antiquity* will be out in October (Burkhardt et al. 1997, 378). Darwin, of course, would have to wait six months to get his shock. There is little correspondence between Lyell and Darwin in 1862, or until Lyell's *Antiquity* appeared in early February 1863, but at any rate the trend is discernible that so long as Darwin thought Lyell was a convert there was no species nominalism talk on his part.

To complete the examination of Darwin's inner circle, we must examine Darwin's correspondence with T.H. Huxley. Huxley does not seem to have come around to evolutionary thinking until shortly before the publication of the *Origin*. Years before this, Darwin confessed his evolutionism in a letter to Huxley (September 2, 1854), stating "I am almost as unorthodox about species as the Vestiges itself, though I hope not *quite* so unphilosophical" (Burkhardt and Smith 1989, 213). This was just after he had read Huxley's savage review of the tenth edition of Chambers' *Vestiges*, which Huxley later recalled as having set him strongly against evolution (Burkhardt and Smith 1989, 213 n. 5). Adrian Desmond (1997) claims that Huxley did an "about-face" (226) in favor of evolution shortly after the last week of April 1856, during which Darwin had Huxley, Hooker, and

Wollaston stay at his home for a weekend. Although Desmond claims that at this stay Darwin did not show his hand to his guests (I don't know how Desmond knows this, but at any rate Hooker and Huxley already knew Darwin's view), the issue of the limits of species mutability was certainly the main topic of discussion, and Desmond argues (in typical fashion) that Huxley converted to evolutionism a few weeks after mainly for political reasons rather than genuinely scientific ones. As Desmond puts it,

> Huxley realized he had been wrong-footed. He was turning, knowing that his secular needs could be well served by transmutation (which could be opposed to supernatural creation), even if it had no jot of "demonstrative evidence in its favour." Could individuals be successively modified? After the Downe weekend he saw no reason why not. After all pigeon breeds differ "as widely from one another as do many species." Darwin had taught him that. [226][15]

Whatever Huxley's main motive, the best evidence suggests that he did not convert until the summer before the publication of the *Origin*. On June 3, 1859, Huxley read a paper to the Royal Institution, with the title "On the Persistent Types of Animal Life." At the end of this paper Huxley (1859a) stated, with regard to the "hypothesis which supposes the species of living beings living at any time to be the result of the gradual modification of pre-existing species," but without any explicit reference to Darwin, that the "hypothesis which though unproven, and sadly damaged by some of its supporters, is yet the only one to which physiology lends any countenance" (153). According to Burkhardt and Smith (1991), "The paper was Huxley's first public defence of CD's theory" (302 n. 3). Prior to the reading of this paper, there is no good evidence (at least that I am aware of) that Huxley thought evolution probable. Instead, he still retained his emphasis on archetypes and the Cuverian purpose of classification, thinking evolution unimportant for classification.[16]

At any rate, what matters, again, is Darwin's *perception* of Huxley's conversion. While working on the *Origin*, Darwin queried Huxley on some points on anatomy. Having received Huxley's reply, Darwin in a letter (March 13, 1859) pointed out to him that "the ideal morphologies of naturalists, are on my notions, real changes in the course of time from one part or organ into another" (Burkhardt and Smith 1991, 262). He then added that he expected Huxley "will be just as savage with me" as he was with Agassiz's *Essay on Classification*, which Darwin himself characterized as "utterly impracticable rubbish." Later that year, close to the publication of the *Origin*, Darwin in a letter to Wallace (November 13) wrote "If I can convert Huxley I shall be content" (375). It should be obvious, then, that prior to the publication of the *Origin* Darwin did not think that Huxley was a convert. Accordingly for my view, prior to Darwin's perception of Huxley's conversion he used the language of species nominalism on him. For example, in response to reading Huxley's "Introductory Essay" in his manuscript later published as *A Catalogue of the Collection of Fossils in the Museum of Practical Geology,* Darwin wrote in a letter to Huxley (December 16, 1857) "I do not understand how you can say that if only fossils existed there would be no difficulty in practically species—as there is variation amongst fossils, as with recent, there seems to be same difficulty in grouping" (Burkhardt and Smith 1990, 506). This example, of course, is not as clear as one would like (and we must remember again that not all of Darwin's correspondence is extant), but it certainly is in the same spirit as Darwin's nominalistic statements with other correspondents. Moreover, we must remember that Darwin knew that Huxley would soon be reading his *Natural Selection,* in which he defines species nominalistically.

Once Darwin was sure that Huxley was a convert, however, his language with regard to species changed. Huxley, having received his presentation copy prior to the official publication date of the *Origin* (November 24, 1859) and having quickly read it, wrote to Darwin (November 23, 1859) "I think you have demonstrated a true cause for the production of species & have thrown the *onus probandi* that species did not arise in the way you suppose on your adversaries." He also dedicated himself to defending Darwin's views, writing "I am sharpening up my claws and beak in readiness" (391). Accordingly, Darwin no longer hints of species nominalism in his correspondence with Huxley. The problem now, however, is that there is barely a hint of species realism either. There is, of course, the problem of lost letters. At any rate, shortly after reading the *Origin* Huxley had asked Darwin to supply him with sources on domestic breeding for a lecture he was preparing in defense of Darwin. In reply, Darwin in a letter (November 27) stated that he knew of no one book and that he relied on direct acquaintance with breeders. A couple of weeks later Darwin supplied in a letter to Huxley (December 13) an enclosure on pigeon breeding. In that letter he included some paraphrases of quotations from the writings of John Matthews Eaton. Although Eaton preferred to refer to domestic pigeons as "species," Darwin in his accompanying letter to Huxley referred to the "Breeds" of pigeons, remarking that if Huxley could see the drawings that Darwin has, he "would have grand display of extremes of diversity" (Burkhardt and Smith 1991, 428). Why would Darwin group all the breeds of pigeons together as one species (as he did in the *Origin*, 20–28), contrary to the breeder Eaton, if he thought that the designation of "species" and "varieties" was arbitrary? Indeed we have seen the answer in chapter 5. Darwin did not designate the different breeds of pigeons as different species—although it would have helped him in his argument with Huxley, as we have seen in chapter 6—because the differences between them were produced by artificial selection and therefore were not really *adaptations*, which for Darwin was what distinguished real species. Darwin's correspondence with Huxley throughout 1860 is preoccupied with reviews and replies to reviews, and there are no more examples to be found for my theory. If Darwin used the language of species realism more explicitly than the example I have provided above, the letters in which he did so are, lamentably, no longer to be found. As such, then, the case of Huxley does not strongly confirm my theory, although there is nothing there to disconfirm it either.

So much for the above four cases of Gray, Hooker, Lyell, and Huxley. I should add that my examination of them is not meant to hide anomalies against my theory. None of the anomalies, however, seem to me terribly difficult. For example, as we have seen in chapter 1, in a letter to Hooker (September 23, 1853) Darwin wrote "Farewell, good luck to your work [on Himalayan species], whether you make the species hold up their heads or hang them down, as long as you don't quite annihilate them or make them quite permanent; it will all be nuts to me" (Burkhardt and Smith 1989, 156). This expression of species realism conflicts with my claim that Hooker did not become a convert until mid-July 1858. However, none of Hooker's letters to Darwin have survived during the almost two years previous to this letter, the previous extant letter being from November 1851 (Burkhardt and Smith 1989, 627). Given the regular correspondence between Hooker and Darwin, then, it is quite possible that in a letter now lost (or in a conversation) Hooker said something to make Darwin think that he had become a convert. Hence, we do not find species nominalism in the letter quoted above.

Another interesting case is a letter written by Darwin to Hooker (April 23, 1861) well after Darwin clearly and rightly considered Hooker a convert. In that letter Darwin wrote "I have been much interested by Bentham's paper in N.H.R; . . . I liked the whole,—all the facts on the nature of close & varying species. . . . I was, also, pleased at his remarks on classification, because it showed me that I wrote truly on this subject in the Origin" (Burkhardt et al. 1994, 99). According to Burkhardt et al. (1994), "Bentham held the view, as did CD, that classification was a matter of convention" (101 n. 7). Thus in this letter Darwin would seem to use the language of species nominalism to Hooker, which goes flat against my theory given the dating of the letter. But if one turns to Bentham's paper (1861)—which was originally read *before* the publication of Darwin's *Origin*—the apparent anomaly quickly disappears. In the very first paragraph Bentham points out that "the whole system of classification depends, in the first instance, on a right understanding of what is meant by species" (133). He then begins by considering common descent in the meaning of "species," and points out some difficulties with that criterion. He then considers an alternative suggestion, which is to eliminate descent and employ similarity (either a common character or set of characters). But he rejects this, because "The species or collection of individuals thus defined, becomes, therefore, as arbitrary as the genus or collection of species, and reduces the rules of classification in the one case, as in the other, to little more than rules of convenience." Bentham then, believing he says "that there exist in nature a certain number of groups of individuals, the limits to whose powers of variation are, under present circumstances, fixed and permanent," proceeds to give his own realist definition of "species" (for botany), which is "*the whole of the individual plants which resemble each other sufficiently to make us conclude that they are all, or MAY HAVE BEEN all, descended from a common parent*" (133). Given Bentham's species realism, then, Darwin's letter to Hooker cannot properly be read as an endorsement of species nominalism.

A more problematic case is Darwin's correspondence with H.C. Watson. Although not normally thought of as part of Darwin's inner circle, Watson was an important correspondent of Darwin's, and Darwin thought very highly of him (cf. Darwin 1859, 48). Interestingly, Watson had become a transmutationist apparently after reading Chambers' *Vestiges*. And although, as we have seen in chapter 1, he shortly afterward came close to embracing species nominalism, he never went the whole way, and continued to make a distinction between book species and real species in nature. We should suspect, then, if my strategy theory is true, that Darwin would never have any reason to employ it against Watson. And yet we find Darwin write in one of his post-*Origin* letters to Watson (July 17, 1861) "The difficulty to know what to call vars & what species,—hopeless—But your suggested plan to take list of forms which some call vars & some species ? print list?" (Burkhardt *et al.* 1994, 207). And we cannot say this is because Watson did not yet accept natural selection as the main mechanism of evolution, because he did. In a letter to Darwin (November 21, 1859) he wrote "Your leading idea will assuredly become recognized as an established truth in science, i.e. 'natural selection'" (Burkhardt and Smith 1991, 385). The problem becomes greatly diminished, however, by two further pieces of evidence. First, the letter that Darwin was replying to is lost. Second and more importantly, the letter by Darwin is not complete, and an important part of it has survived in a copy of that part made by Watson (Burkhardt *et al.* 1994, 208 n. 5). Evidently the issue was only over "very closely allied forms," more specifically "how often do such allied forms (held species by some, vars. by other botanists) live mingled in same spot,—how often in same district, but in different stations,—and how often in different geographical districts." Since, as we have seen in chapters 1 and 3, Darwin did not believe that messy situations were the norm, this letter, then, should not be read as an expression of species nominalism.

Another interesting case, and it is the final one I shall examine, is Darwin's correspondence with A.R. Wallace. In what is his earliest extant letter to Wallace (May 1, 1857), Darwin wrote "This summer will make the 20th year (!) since I opened my first-note-book, on the question how & in what way species & varieties differ from each other" (Burkhardt and Smith 1990, 387). In this letter there is no hint of species nominalism. And yet Darwin did not yet think of Wallace as an evolutionist. Wallace had written a paper on geographical distribution (Wallace 1855), in which he claimed it a law that "Every species has come into existence coincident both in space and time with a pre-existing closely allied species" (186, 196). In the margin of his copy, Darwin wrote "nothing very new," "Uses my simile of tree—It seems all creation with him," "I sh^d. state that put generation for creation & I quite agree" (Burkhardt and Smith 1989, 522 n. 1). It is not known when Darwin wrote these words in the margin, but Wallace's article was favorably referred to by Edward Blyth in a letter to Darwin of the same year (December 8, 1855) (Burkhardt and Smith 1989, 519, 520). In fact, two years later, after an exchange of a number of letters with Wallace (Burkhardt and Smith 1990, 388 n. 2, 457), Darwin still thought of Wallace as a creationist. In a letter to Wallace (December 22, 1857) he says "But you must not suppose that your paper [Wallace 1855] has not been attended to: two very good men, Sir C. Lyell & Mr E. Blyth at Calcutta specially called my attention to it. Though agreeing with you on your conclusion in that paper, I believe I go much further than you" (Burkhardt and Smith 1990, 514).[17] Wallace, unknown to Darwin, had become an evolutionist in 1845 shortly after reading Chambers' *Vestiges*. And of course many years later, in February of 1858, while ill with a malarial fever in the Malay archipelago, Wallace independently conceived of evolution by the mechanism of natural selection. In the same month he wrote his seminal paper (Wallace 1858) and sent it to Darwin in the hope that Darwin would send it to Lyell, which unknown to

Wallace forestalled Darwin (Kottler 1985, 368–371). Darwin received Wallace's letter (which is no longer extant) with the accompanying article on June 18, 1858 (Burkhardt and Smith 1991, 107, cf. xvii–xviii) and was of course shocked. What is important for my purposes is that Darwin could no longer think of Wallace as a creationist. So why then, if my theory is correct, did Darwin not employ his nominalism/realism strategy on Wallace? The answer, it seems to me, is because prior to late June 1858—which is a little after Darwin got his shock and it was quickly resolved that papers by Darwin and Wallace would be read to the Linnean Society on July 1; cf. Burkhardt and Smith (1991, 121 n. 3)—Darwin was very selective about who he revealed his evolutionism to. In fact, the list does not go much beyond Hooker, Lyell, Gray, Huxley, and Watson, in which cases, as we have seen, my strategy theory applies very well. Wallace, being principally a specimen collector and in Darwin's eyes (prior to June 18, 1858) an amateur theorist, simply didn't rate (cf. Endersby 2003, 397–398). Therefore we should not be surprised to find that my strategy theory does not apply to the correspondence between Darwin and Wallace.

In sum, I think it should be evident from the above examples, including even the difficult ones, that there is statistically significant evidence supporting my theory that Darwin in his correspondence took a stance of species nominalism only to replace it with a stance of species realism once the desired conversion, in a minimal sense, had been accomplished. Contrary to Beatty's thesis, there is no reason to believe that Darwin did this in order to better communicate his theory of evolution. On the contrary, it is abundantly evident with the group that would eventually become his inner circle—namely, Hooker, Lyell, Huxley, and Gray—that they knew perfectly well what he was talking about, even though they might have disagreed with him at the time. Instead, Darwin's strategy was a further part of his argumentation. In league with his arguments about artificial selection, biogeography, classification, and so forth, his *reductio ad absurdum* strategy was designed to loosen his correspondents from their orthodox position on species. And it would serve the same function in his published writings. In short, if they believe that varieties are not real and if Darwin can demonstrate to them that varieties are incipient species, then they must conclude that species are not real either. Otherwise they must change their concept of both varieties and species.

With all of this in mind, it is now time to widen our scope and look at theories on the nature of concept change in scientific revolutions. The case of Darwin on species and varieties, as now newly understood from this and previous chapters, will prove to be an interesting case study.

Chapter 9

Concept Change in Scientific Revolutions

The case of Darwin on the nature of species presents a major example of concept change during a scientific revolution. It becomes even more interesting if we add to it the species problem from the time of Darwin to the present. At any rate, if we compare scientific revolutions to mass extinctions, the Darwinian revolution must rank among the top five, possibly as number one, comparable to the Permian extinction. Of course, just as with mass extinctions, it would be a mistake to generalize from one case, no matter how important that case. Nevertheless, it is interesting and instructive to compare the case of species, both in the case of Darwin and in the case of biology from the time of Darwin to the present, with a number of prominent theories on concept change in scientific revolutions. In particular, it will be interesting to see how preconceived theories, drawn from other examples in science, have been applied to the case of Darwin, or how they would apply, and how they must be revised because of the analysis presented in the previous chapters.

Perhaps a good place to begin is with the pragmatist philosopher John Dewey's 1909 essay on the influence of Darwin on philosophy, published in Dewey (1910). Dewey argues that the concept of *eidos*, developed by the ancient Greeks and termed *species* by the scholastics, was a concept of knowledge right up to the time of Darwin. Species were "the sacred ark of permanency" (1) in the "flux" of change (5), each of them a "fixed form and final cause, . . . the central principle of knowledge as well as of nature" (6).

Natural selection, on the other hand, "cut straight under this philosophy" (11), such that the *Origin* constituted an "intellectual face-about" (3), a "protest" against "the long-dominant idea" (4), and as such "precipitated a crisis" (2).

What is especially interesting about Dewey's (1910) argument is his implied claim about the future of the concept of *eidos/species*. He says "Old ideas give way slowly . . . . We do not solve them ['old questions']; we get over them. . . . Doubtless the greatest dissolvent in contemporary thought of old questions . . . is the one effected by the scientific revolution that found its climax in the 'Origin of Species'" (19).

Part of what would make Dewey's argument attractive is the occurrence of overt species nominalism in Darwin's *Origin*, as we have seen in chapter 1 and which Dewey was surely aware of. But another part is contemporary with Dewey, namely, the surge in species nominalism in biology around the time of his writing (cf. Morgan 1903, 33; Bessey 1908, 218; Arthur 1908, 244, 248; Cowles 1908, 267; Coulter 1908, 272).

Subsequent history, however, has not been kind to Dewey's argument. There are some today, of course, who advocate species nominalism, but modern biology has experienced nothing like the surge in species nominalism around the time of Dewey (1910). Instead, since that time, species nominalism has constituted a minority view compared with species monism and species pluralism. Indeed biology since the forging of the Modern Synthesis has experienced a bush pattern proliferation of species concepts (Stamos 2003).

To be sure, there are some today, such as Robert O'Hara (1993), who have urged that we should try "not to solve the species problem, but rather to get over it" (232). For one, I don't see this happening. The species problem is not like a failed marriage. We cannot hope to get over it by trying to forget about it and get on with our lives. For a start, biology needs a common currency, where species constitute the "trade of ideas" in biology (Paterson 1985, 137; cf. Maynard Smith 1975, 217–218). When we encounter an organism, whether old or new, we want to know "What is it?" This is not just of theoretical but of practical importance, as in biological control and conservation biology. "Correct identification," as David Rosen (1978) puts it, "provides the key to any available information about the species, its distribution, biology, habits, and possible means of control, which would otherwise have to be independently investigated at a considerable expense of time and effort" (24).

Accordingly, most subdisciplines in biology require a species concept in order to convey much of their information. Moreover, species are the basal units in taxonomy, and it is difficult to imagine taxonomy without them.

The problem then remains, as Cracraft (1989) puts it, that "Different species concepts organize the world differently and conflictingly" (40). Ignoring this problem will not solve it or make it go away. On the contrary, with the growing environmental crisis the pressure on biology to solve the species problem has become greater than ever (Cracraft 1997, 332–337). The past 95 years, since the time of Dewey's writing (1910), has borne this need out.

In the case of species, then, Dewey did not get it right on the nature of concept change during and after a revolution. Much more to the point would seem to be Thomas Kuhn's position, beginning with his now classic *The Structure of Scientific Revolutions* (1970). Against the attempt to interpret his work as entailing kind nominalism, Kuhn later (1993) affirmed that he had always held that a scientific revolution will require not simply individuals but "both kind-concepts and their names" (316). Back in *Structure* (1970), Kuhn claimed that during normal science scientists learn to group objects into "similarity sets." During and as a result of a scientific revolution, however, "some of the similarity relations change. Objects that were grouped in the same set before are grouped in different ones afterward and vice versa" (200). The classic example Kuhn gives is that of the concept of planet during and after the Copernican revolution, where the concept did not become nominalistic but instead the extension of the concept was changed to include the earth and exclude the sun. One can also think of the concepts of sun and star, for which Giordano Bruno was burned at the stake for maintaining that each star is a sun and our sun is a star. Kuhn goes on to say that "Since most objects within even the altered sets continue to be grouped together, the names of the sets are usually preserved. Nevertheless, the transfer of a subset is ordinarily part of a critical change in the network of interrelations between them" (200).

I shall return to Kuhn's last sentence above shortly. As for what he says before it, he is pretty much on when it comes to the case of Darwin. As we have seen, Darwin did not really attempt to eradicate the concept of species. But he did not keep the extension of species names the same either. Instead, in a significant number of cases, but certainly not most, he changed the extension of species names in accordance with his implicit evolutionary species concept, which was in accordance with the *vera causa* ideal of his fellow naturalists for both scientific explanation and natural classification. Moreover, in line with what Kuhn says above, Darwin generally preserved the names of species. He did not invent new species names for already known organisms. (This was true even in the case of the taxonomic mess that confronted him in his work on barnacles.) Instead, in a number of cases what were called "varieties" of one species were now called "species" by Darwin (as with primroses and cowslips), while in a number of cases what were singularly called "species" were broken into a number of different species (as with domestic dogs and cattle).

In all of this we can see at work what Jagdish Hattiangadi (1987) calls the progressive and conservative forces in language change. The progressive force is the increase in knowledge. Keeping to what is relevant to the case of Darwin and species, kind concepts and categories are theories about the world, and they necessarily need to change if they become untenable in the light of new knowledge. In the case of Darwin, the progressive force was the evidence for evolution by natural selection and all that this involved. The conservative force, on the other hand, is the need for communication. As Hattiangadi puts it, "So long as we try to communicate a new idea, it is necessary to be modest in our linguistic demands—otherwise we will be misunderstood" (172). For Darwin this meant that he could not afford to get rid of species talk. His fellow language users were his fellow naturalists and the public at large. Consequently he kept the term "species" and retained in most cases, but not all, the use of that term.

Returning to Kuhn (1970), what is problematic in his claim above is the last sentence that I said I would return to. It refers to his famous theory of the "incommensurability" (incomparability) of competing paradigms during a scientific revolution. According to Kuhn, "when such redistributions [in the sets or kinds discussed above] occur, two men whose discourse had previously proceeded with apparently full understanding may suddenly find themselves responding to the same stimulus with incompatible descriptions and generalizations." For Kuhn, such problems are not merely verbal, so that "they cannot be resolved simply by stipulating the definitions of troublesome terms" (201). Kuhn would later (1993) call incommensurability "the central innovation introduced by the book [*Structure*]" (315), but to understand it better we need to fill it out with some of Kuhn's other key concepts. For Kuhn (1970), the history of a science is characterized mainly by normal science—comparable, I might add, to the period of stasis for the history of a species in the theory of punctuated equilibria—during which puzzles are solved within the framework of the existing paradigm and it is the abilities of the scientists themselves that are constantly tested, not the paradigm itself. If failures or anomalies start accumulating, however, the situation might eventually reach a threshold or crisis point, where the abilities of the scientists working under the paradigm cease to be questioned and the paradigm itself becomes questioned. This period of crisis "generally" (67) precipitates a scientific revolution[1]—comparable again to the punctuation in punctuated equilibria—where "The novel theory seems a direct response to crisis" (75). Kuhn added to this his theory of incommensurability, which he characterizes as involving (i) different kinds, definitions, and problems (148), (ii) misunderstanding, "communication breakdown" (149, 201), even "talking through each other (132), and (iii) "different worlds" (111) in the sense of radically different worldviews, such that the shift in an individual scientist from one paradigm to another during a scientific revolution, if that shift occurs at all (Kuhn subscribed to the view that quite often the older scientists simply die off), "cannot be made a step at a time" but is instead sudden, being much like the gestalt switch in psychology (150), as with the famous duck/rabbit (111), or like a religious conversion experience (151), involving "faith" that the new paradigm will succeed in solving problems and is on the right track (158).[2] Granted, Kuhn does qualify his meaning such that he is not to be taken as saying "no arguments are relevant or that scientists cannot be persuaded to change their minds" (152). Nevertheless, he argues (199–200), especially in Kuhn (1977), that the underlying reasons will be largely psychological and social (139, 325), the values that guide conversion often shifting and conflicting (322–325), and he never calls them epistemic. Moreover, he explicitly (1970) disavows that successive scientific revolutions approach nearer to the truth (170–173, 205–207), although he allows for progress in an instrumentalist sense within a paradigm (166).[3] In either case, gestalt switch or conversion, what is involved, says Kuhn, is "neural reprogramming" (204). He also characterizes incommensurablity as a matter of speaking different languages (this became his exclusive meaning in later writings), so that translation, which for him is never more than partial, is therefore required, which in turn serves a powerful role in conversion (202). Although Kuhn can be found claiming that a newly established paradigm is not only "incompatible" with the one it replaced but "often actually incommensurable" (103), where the keyword is "often," he can later be found saying (Kuhn 1977) that "communication between proponents of different theories is inevitably partial," such that these limits "make it difficult or, more likely, impossible for an individual to hold both theories in mind together and compare them point by point with each other and with nature" (338). Moreover, if a scientist makes the change it is like the change from being a native speaker of one language to being like a native speaker of another (339). Thus, all things considered, for Kuhn "an individual's transfer of allegiance from theory to theory is often better described as conversion than as choice" (338), the word "choice" being inappropriate partly because it depends on comparison (339).

Kuhn in his analysis of scientific revolutions notoriously underrepresented biology (cf. Hoyningen-Huene 1993, 5). What is especially interesting about the Darwinian revolution, in particular Darwin and the concept of species, is how poorly it fits Kuhn's model.

For a start, Frank Sulloway (1996, 16–19, 346–348) argues that the Darwinian revolution does not seem to have been precipitated by a crisis. Indeed, says Sulloway, "the more radical the revolution, the less it seems to accord with Kuhn's formula" (347). Interestingly, Kuhn (1977) tried to make out a crisis in the decades before the *Origin*, vaguely referring to

> fields like stratigraphy and paleontology, the geographical study of plant and animal distribution, and the increasing success of classificatory systems which substituted morphological resemblances for Linnaeus's parallelisms of function. The men who, in developing natural systems of classification, first spoke of tendrils as "aborted" leaves or who accounted for the different number of ovaries in closely related plant species by referring to the "adherence" in one species of organs separate in the other were not evolutionists by any means. But without their work, Darwin's *Origin* could not have achieved either its final form or its impact on the scientific and the lay public. [139–140]

In all of this it is hard to see a crisis. The use of metaphorical language, such as "aborted" leaves, hardly gives Kuhn what he wants. Sulloway would seem to be right when he argues that there was no buildup in anomalies or failures in puzzle solving. Seemingly anomalous findings, such as new fossils found in the fossil record or Darwin's Galapagos discoveries, were rather easily accommodated into the existing explanatory framework. Moreover, from the time before Lamarck to the *Origin*, "evolutionary thinking was considered a sign of scientific incompetence rather than a solution to widespread problems in interpreting the existing data" (17). I might add that, as we have seen in Chapter 1 in the quotation from Wollaston, special creation was considered almost an axiom, while, as we have seen in chapters 2 and 7, limited variation and reversion were considered to be laws. This did not really begin to change until the *Origin* was published. Indeed, as Sulloway further remarks, "Chambers, Darwin, and Wallace jointly *created a crisis where none had been before*" (347).

Turning now to incommensurability, it might appear that Darwin himself was aware of the problem of communication breakdown and complained of it in his efforts to gain converts. For example, in a letter to David Thomas Ansted (October 27, 1860), in particular with respect to reviewers of the *Origin*, Darwin wrote "I declare that the majority of readers seem utterly incapable of comprehending my long argument. . . . I am often in despair in making the generality of naturalists even comprehend me. Intelligent men who are not naturalists and have not a bigoted idea of the term species, show more clearness of mind" (Burkhardt *et al.* 1993, 446). And yet, as Sulloway (1996) argues,

> Darwin himself seems to have considered evolution as *fully commensurable* with creationism. His *Origin of Species* was "one long argument" comparing how well the available biological evidence could be interpreted by creationism and evolution. He sought to demonstrate, point by point, that rational criteria consistently favored the evolutionary alternative. . . . Darwin was so successful in showing how commensurable evolution and creationism were that he forced his readers to make a direct choice between the two theories. Most reasonable people had to admit that evolution was the superior theory. After Darwin's careful dissection of the Creator's handiwork, the Creator began to look so mindless that many creationists wisely sought to move Him into the biological background as a "first cause." [349–350]

I would add to this that Darwin's "one long argument" is an excellent example of what has come to be known in philosophy of science as *inference to the best explanation*. This is a model of scientific explanation that in many ways is superior to the traditional deductive-nomological model, one reason being that in biology explanation is not, if ever, in terms of laws of nature (cf. Mayr 1982, 37; Stamos 2006). At any rate, at the very heart of inference to the best explanation is *contrastive explanation*. To use a simple example given by Peter Lipton (1990), it is not enough to explain that Kate won the essay prize because her essay was excellent. A full explanation would have to include why it was better than the essay of the other finalist, Frank. As Lipton puts it, "One reason that explaining a contrast is sometimes harder than explaining the fact alone is that explaining a contrast requires giving causal information that distinguishes the fact from the foil, and information that we accept as an explanation of the fact alone may not do this" (252).

Sulloway goes on to argue that firstborns, those who were actually the oldest among their siblings or functionally took their role, would be more likely to claim incommensurability, since firstborns tend to be more conservative and less open to new ideas than laterborns, who tend to be more flexible, even rebels. As Sulloway (1996) puts it, "Psychologically speaking, incommensurability is a problem for inflexible people, not for those who are open to experience" (349). This is of a piece with Sulloway's enormous statistical analysis, 26 years in the making, based on the biographies of scientists involved in scientific revolutions, in particular the Copernican, Newtonian, Darwinian, and Einsteinian revolutions. Sulloway found, remarkably, that those scientists who defended the traditional paradigm tended to be firstborns, while those who defended the emerging new paradigm tended to be laterborns. Rationality, thus, says Sulloway, is "typically subject to a *threshold* effect" (341), laterborns having a lower threshold and therefore tending more quickly to see the better theory for what it is, and firstborns having a higher threshold. Indeed as empirical evidence and argument accumulate, and the revolution comes to a close, birth order differences tend to diminish (337), the firstborns either dying in their stubbornness or converting to the new science (34–36).[4]

What is magnificent about Sulloway's (1996) work is that he literally spares no effort, not only in developing theories but in subjecting them to rigorous testing. This latter part is terribly missing in the humanities. For example, feminists and sociologists seem averse to the point of hostility in subjecting their theories to tests, whether to the statistics of rape (which point away from culture and environment being the sole cause), to the statistics of homosexuality (which in males increasingly points to a partly evolutionary/genetic cause), or to the statistics of behavioral differences between men and women (which fit an increasingly evolutionary view). Marxists too seem bent on believing, no matter how much evidence to the contrary, that human nature is completely plastic. Social constructionist history of science (which I shall consider in the next chapter) fares no better.

At any rate, returning to the topic of incommensurability, a major part of the problem created by Kuhn stems from his adherence to the traditional theory of meaning (cf. Kuhn 1970, 198), according to which the intension (concept, definition, description, sense, connotation) of a term determines its extension (reference, denotation), either by a conjunction of defining predicates (e.g., Mill, Frege, Russell) or by a cluster of defining predicates (e.g., Wittgenstein, Searle). This was a view that Kuhn maintained throughout his career and refused to change (cf. Kuhn 1993, 316). On this view of meaning, given that a revolutionary theory (the new paradigm threatening to displace the old) typically employs radically different meanings for at least some of the key terms, it would happen that scientists on opposite sides of the revolutionary divide would be referring to different kinds of things even though they would be using the same terms. Consequently they would be having a communication breakdown. Moreover it would be impossible to say that knowledge of kinds progressed over time, from the traditional paradigm to the revolutionary paradigm that displaced it. Instead there would simply be different theories of kinds, with an overlap in terminology but not an overlap in reference.

Many philosophers (cf. LaPorte 2004, 117–118 for references) have found that the incommensurability thesis of Kuhn is undermined by the new causal theory of reference. This theory, connected principally with Saul Kripke, Hilary Putnam, and Keith Donnellan (cf. the Introduction and papers in Schwartz 1977), is arguably superior to the traditional theory as both a theory of meaning for proper names and for natural kinds. Beginning with proper names, for which the causal theory was originally designed, it is fundamental to the traditional theory of meaning that a set of defining predicates determines the reference of, say, "Aristotle," such as "son of Nicomachus and Phaestis," "student of Plato," and "teacher of Alexander the Great." However, if these defining predicates (which the proper name represents) turn out to be historically false, then the name "Aristotle" has no historical reference and we are talking about no one. This is a paradoxical result that the causal theory avoids. On this theory, what fixes the reference of the name "Aristotle" is an initial baptism, an ostension, typically accompanied by a description that aids in the ostension, not in the sense of listing attributes but in fixing the reference, such as, "This baby is 'Aristotle.'" Consequently, references to Aristotle by later generations can correctly refer to the historical person, Aristotle, even if everything they believe about him happens to be historically false, just so long as their use of the name "Aristotle" is part of a causal/historical chain leading back to the initial baptism. For the creators of this new theory of meaning, the reference of natural kind terms (e.g., "gold," "water," and "tiger") are fixed in basically the same way, by referring to a sample of the kind and typically using a reference-fixing description.[5]

The way in which this theory of meaning undermines incommensurability is that scientists on opposite sides of the revolutionary divide can be talking about the very same kinds even though their theories about those kinds are radically different. Their meaning, in the sense of reference, can be exactly the same. Realism about kinds can therefore be maintained.[6] Moreover, progress in knowledge about the kinds can be accomplished because the reference between theories remains the same.[7]

Indeed it is this theory of meaning that forms the background to Beatty's (1985) theory about Darwin on "species."[8] Again, as we have seen in previous chapters, according to Beatty (1985) Darwin "tried to get beyond definitions to referents" (269), "he used the term ['species'] in accordance with *examples* of its referential use by members of his naturalist community" (277), "By trying to talk about the same things that his contemporaries were talking about, he hoped that his language would conform satisfactorily enough for him to communicate his position to them" (266). Here we can see Beatty claim, in other words, that Darwin surmounted the incommensurability problem, that he bridged the revolutionary divide and facilitated communication, by retaining the reference to species made by his fellow naturalists. And indeed Philip Kitcher (1993), explicitly following Beatty (1985)—ironically whose work he himself influenced (cf. note 8 above) with his (Kitcher 1978) discussion on and modification of the causal theory of reference—makes explicit what Beatty left more or less implicit: "the term ['species'] retained its old referent (the set of species taxa after Darwin is the same). Similarly, there is continuity of reference of the term 'homology' (and its correlative, 'analogy')" (32). Moreover, "Partly because they were talking about the same groups of species and the same traits as homologous, partly because the loci of disagreement were so readily identifiable, Darwin and his critics had not the slightest difficulty in communicating their ideas to one another" (32 n. 46).

Interestingly, the case of Darwin on species actually fits much better, though not perfectly, Kitcher's own modification of the causal theory of reference, which has at its core what he calls "reference potential." As Kitcher (1993) puts it, the reference potential of a scientific term is "the compendium of modes of reference for a term (type) . . . . reference potentials are typically heterogeneous: that the linguistic community to which a scientist belongs allows a number of distinct ways of fixing the reference of tokens of terms" (78). Kitcher (1978) is where Kitcher first developed this theory, and his focus there is on the concept of phlogiston. Although the reference for this term was first fixed in a particular way, namely, the reference-fixing description (possibly by Stahl) that phlogiston is "that which is emitted in all cases of combustion," the reference later expanded, in a way that all subscribers to the theory of phlogiston agreed upon (540), to include also "dephlogisticated air," what they called the gas obtained by heating the red calx of mercury, and which Lavoisier later came to call "oxygen." This meant that even though "phlogiston" failed in fact to refer, its related concept "dephlogisticated air" did not. Equally important, it meant that the event types responsible for each token of the word "phlogiston" by the community that subscribed to phlogiston theory became a set of two types, or reference-fixing descriptions. (I am ignoring a third type, "inflammable air," for the sake of simplicity.) In this way, according to Kitcher, the reference potential of a term shifts, sometimes expanding in terms of the set of event types or reference-fixing descriptions, sometimes contracting. This allows us to avoid "conceptual relativism" (535), to "allow for radical conceptual revision without conceptual discontinuity" (544), and to "clarify the idea that theoretical concepts must absorb theoretical hypotheses and so enhance our understanding of conceptual change in science" (543).

What is interesting about this theory is how well it applies to Darwin, in the case of "species" but even more importantly in the case of "variety." Although it is virtually impossible (as far as I know) to determine who began the first reference-fixing description for the term "variety," by the time of Darwin the term clearly had a heterogeneous reference potential, as we have seen in Chapter 7. The difference between the case of "phlogiston" and the case of "variety" is that not all naturalists at the time of Darwin, or from the time of Linnaeus to Darwin, agreed upon the total set of reference-fixing descriptions, or token-initiating events, for the term "variety." (Indeed the reference of the term "variety" was no more fixed than the reference for the term "species.") And yet this does not seem to matter. The important fact is that the heterogeneous reference potential was there. What Darwin did, again as we have seen in chapter 7, was attempt to contract that reference potential, limiting the use of the term to only what would qualify as an "incipient species," which had to be a group of organisms characterized by a heritable trait that either is being evolved or could be evolved by natural selection into an adaptation. The heterogeneous set of reference-fixing potentials for the term "variety" was therefore in an important respect similar to the set for the term "phlogiston," in that, much like Lavoisier with what would later be called "oxygen," Darwin established continuity with one item in that set, namely, those varieties called as such by his fellow naturalists that qualified as incipient species. This also explains why and how Darwin attempted to shift the reference potential of the term "species." Even though his fellow naturalists were mistaken in their theory-laden heterogeneous set of reference-fixing descriptions, many of the entities they referred to as "species" were in fact species in accordance with Darwin's new reference-fixing description, namely, his implicit species concept defined near the end of chapter 6. Of course, unlike Lavoisier and the reference of the term "oxygen," the shift in the reference potential of the term "species" that Darwin tried to effect did not become fixed. Instead, the term is still to this very day shifting between different reference-fixing descriptions, with no agreement or consensus among the expert users of the term (cf. Stamos 2003), even as to whether species are evolutionary (89, 323). The situation in modern biology is thus very different than the situation in modern chemistry.

Kitcher, of course, missed all of this, apparently for the simple reason that he followed Beatty (1985) in thinking that Darwin retained the reference of the term "species" of his fellow naturalists. He consequently thought that there was no incommensurability problem. For others who subscribe to the causal theory of reference, however, and also to Beatty's theory on Darwin, the matter is not so simple. Joseph LaPorte (2004), for example, accepts a modified version of the causal theory (5–7), and he also explicitly accepts Beatty's (1985) claim that Darwin retained the reference of the word "species" of his fellow naturalists (192 n. 13). But unlike Kitcher, LaPorte argues that when applied to the case of the Darwinian revolution Kuhn was partly right about incommensurability. He refers to evidence such as Darwin's letter to Ansted quoted earlier in this chapter, in which Darwin seems to complain about communication breakdown, as well as Wollaston's (1860) claim in his review of the *Origin* that "it is no sign of metaphysical clearness when our author [Darwin] refuses to acknowledge any kind of difference between 'genera,' 'species,' and 'varieties,' except one of *degree*" (133), in which Wollaston seems to reject Darwin's view as being too paradoxical. Thus, for LaPorte, Kuhn's incommensurability thesis when applied to the concept of species in the Darwinian revolution "turns out to enjoy more merit than a first glance suggests" (122). For LaPorte, Kuhn is partly right not because of the traditional theory of meaning but because the reference-fixing descriptions of Darwin's fellow naturalists were not only individually vague but jointly conflicting. For example, for one (William Hopkins) evolution destroys the reality of species, while for another (Henry Fawcett) evolution means that the whole tree of life is one or a few species (123–126). What Darwin did, then, according to LaPorte, is retain their reference for "species" but replace their vague and conflicting reference-fixing descriptions with a new one that was clearer and more knowledgeable about what they were in fact referring to. Indeed, "Darwin understood more than did speakers before him about both what he called 'species' and what speakers before him called 'species'" (129), so that his new use of "species" constituted progress over the previous uses of "species," progress in the sense of "replacing vague statements that are neither clearly true nor clearly false with straightforwardly true statements" (131).

Not surprisingly LaPorte is quite vague on the content of Darwin's "newly refined use of 'species'" (123), except to say that "For Darwin the supposition that there are lots of 'species' is retained, and the supposition that all blood relatives of a lineage's original population belong to that species is discarded" (125). Of course another part of Darwin's "newly refined use of 'species'" is that statements such as "New species arise by evolution" are true (131).

Failing to get even close to what Darwin really meant by "species," however, is not the main problem with LaPorte's account, although it is a big one. Much deeper is that he follows Beatty (1985) on Darwin's referential use of the word "species." The problem is enormous because, as we have seen mainly in chapter 5, Beatty's thesis is highly inaccurate. Indeed, it would appear that Beatty took the causal theory of reference and attempted to *impose* it on the evidence, rather than test the theory against the evidence. Remarkably this quickly and easily became the received view. Rather than look at Beatty's claim as a theory that needed to be tested against the evidence, it was simply accepted as a fact. At any rate, Darwin, as we have seen, did not in fact simply retain the reference of the term "species" of his fellow naturalists. Instead, he quite often changed that reference. Because Beatty missed this, he failed to see what Darwin was really doing. Those who followed Beatty not surprisingly failed to see it too. In not seeing that Darwin did not always follow the referential use of the term "species" of his fellow naturalists, they saw no reason to inquire any further. But once one sees that Darwin often changed the reference of the term "species" of his fellow naturalists, one is then naturally led to inquire as to why. That question led to the previous chapters culminating in the reconstruction of Darwin's species concept in chapter 6, the implicit but highly developed species concept that Darwin had in the back of his mind every single time he changed the reference of the term "species." All of this requires us to look anew at the incommensurability problem as applied to the Darwinian revolution.

LaPorte indeed tries to retain some of Kuhn's incommensurability thesis as applied to Darwin and the Darwinian revolution, but because he follows Beatty, he has failed to see what the real situation was, and consequently what the real problem was. For a start, it is surely significant that Darwin held back his true meaning of "species" and defined the term instead nominalistically (for both category and taxa). This certainly adds an interesting twist to communication breakdown. Indeed, the breakdown on this topic has continued to the present day! But this is not to say that Darwin's fellow naturalists would not have understood what he meant by "species" had he taken the time to define it (assuming for the sake of argument that my reconstruction in previous chapters is essentially correct). The problem was not simply one of language or meaning or communication. Nor was it that the species concept of Darwin's fellow naturalists was "theory laden," as Beatty (1985, 271, 279) has claimed. That is to put it much too mildly. The problem was not theories but *deep-seated beliefs*. Darwin, in the letter to Ansted quoted above, refers to the "*bigoted idea* of the term species" (italics mine). This compares favorably (in spite of my analysis in chapter 8) with what Darwin wrote in a letter to Asa Gray (December 21, 1859) with regard to an American reprint of the *Origin*, that "I have made up my mind to be well abused; but I think it of importance that my notions sh^d. be read by intelligent men, accustomed to scientific argument though *not* naturalists. It may seem absurd but I think such men will drag after them those naturalists, who have *too firmly* fixed in their heads that a species is an entity" (Burkhardt and Smith 1991, 440, latter italics mine). The problem was not meaning or theories but beliefs, *deep-seated* beliefs shared by naturalists but not by other intelligent men, namely, the "axiom" of special creation as we have seen Wollaston call it in chapter 1, the "assumption" or "faith in certain impassable barrier between existent species" as we have seen Watson call it also in chapter 1, the belief in the sterility barrier as we have seen in chapter 6, and the belief in the laws of limited variation and reversion as we have seen in chapters 2 and 7.

These and other beliefs did not pose a barrier to understanding and therefore to choice. For example, not everyone believed in the sterility barrier, and yet those who did believe in it knew what those who did not were talking about and *vice versa*. Similarly, lumpers knew what splitters were talking about and *vice versa*. In each case the problem was not understanding but agreement.

The real problem, then, was not one of content or meaning, but of prejudice and dogma. What further added to this were the religious and moral dimensions. Adam Sedgwick, for example, in a letter to Darwin (November 24, 1859), was surely not speaking only for himself when he wrote "Passages in your book . . . greatly shocked my moral taste" (Burkhardt and Smith 1991, 397). In line with this, Darwin was surely not simply joking when he referred to the conversion to his views as "*perversion*" (e.g., Burkhardt and Smith 1991, 241, 347), a usage originally given to him by Lyell (349).

In the case of Darwin, then, we have every reason to believe that his fellow naturalists would have sufficiently *understood* his real species concept had he defined it, and therefore would have been able to make comparisons between his concept and other species concepts. This is because every element in Darwin's species concept would have been perfectly familiar to them. As Loren Eiseley (1958) in a different context put it, "The complex of ideas which later went to make up Darwinism was widely enough diffused in the eighteenth century" (119). Moreover, no mental acrobatics would have been needed to understand Darwin's unique combination of the elements, so very much unlike Einstein's concept of curved space (one of Kuhn's favorite examples of conceptual incommensurability; cf. Kuhn 1970, 149). For a start, the importance of adaptations was shared by Darwin's fellow naturalists, as was the importance of common descent (which they limited to within a species). The horizontal dimension would have been understood from Darwin's language analogy. The process that Darwin claimed makes species, namely, natural selection, would have been well understood because they knew of artificial selection and it was well understood. They also would have appreciated the fact that Darwin's species concept was fully in accord with their *vera causa* ideal for both scientific explanation and classification, as we have seen in chapter 5. Moreover, many of the species of Darwin's fellow naturalists coincided in terms of reference (restricted of course to contemporary organisms, whether living or fossil) with the species determined by Darwin's species concept, since the species of the former would have had to have at least one unique adaptation (as they often did) if there was to be overlap. And of course no one doubted that species (their members) have adaptations. That Darwin did not provide his definition of "species," then, was not because of a concern for communication. Indeed this was not much of a problem in itself and Darwin knew it, which is revealed in the fact that he was not afraid to change the reference of what his fellow naturalists called "species," again as we have seen in chapter 5. Instead Darwin's concern was for *conversion*, and it was because of this concern that he employed not only the nature of his "one long argument" in the *Origin* but also the nominalistic strategy revealed in chapter 8.

The subject of conversion presents a further problem for Kuhn's model. Kuhn's claim about conversion is a psychological one, that conversion in an individual scientist during a revolution is saltational not gradual. This follows from his connection of conversion with a gestalt switch, irregardless of whether he further thought of conversion in the religious sense (cf. note 2 above). This is a theory that is testable. The evidence that constitutes the testing is, of course, indirect, both because we are talking about the minds of scientists and because we are dealing with history, where most of the principals are dead. In many ways the situation is analogous to the theory of punctuated equilibria. In both cases (confining Kuhn's thesis to the Darwinian revolution) the organisms in question are all dead and we are dealing exclusively with historical records, so that the evidence is indirect. Moreover, the debate is between saltation versus gradualism (of course with paleontology the "saltation" is in terms of geological time, which according to punctuated equilibria is 5,000–50,000 years). Finally, the issue is one of frequency (cf. Stamos 2003, 222–223). Interestingly, in the case of the theory of punctuated equilibria the evidence does not seem to be in its favor (cf. Erwin and Anstey 1995). In the case of Kuhn's theory of conversion when applied to the Darwinian revolution, the evidence against seems even stronger.

To begin with, there is certainly indirect evidence that supports Kuhn's theory. For example, in his autobiography, when looking back at the revolution he inaugurated, Darwin (1876a) appears to view conversion as a kind of gestalt switch. In an often-quoted passage he says "What I believe was strictly true is that innumerable well-observed facts were stored in the minds of naturalists, ready to take their proper places as soon as any theory which would receive them was sufficiently explained" (124).

When we look more closely at the Darwinian revolution, however, we find time after time that among key cases the "conversion"—as we have seen in chapter 8 the term was used quite frequently by Darwin and his contemporaries—was not saltational but gradual, no matter whether the conversion was full or partial. We have seen this in chapter 8 in the case of Lyell, in the case of Gray, in the case of Huxley (which was even longer if Di Gregorio 1982 is correct), and in the case of Hooker. Indeed Hooker's conversion, as we have seen, was extremely gradual, and this is borne out by Henry Fawcett (1860) in his review of the *Origin*, in which he quotes from a speech Hooker gave at a then recent meeting of the British Association: "I knew of this theory [Darwin's] fifteen years ago; I was then entirely opposed to it; I argued against it again and again; but since then I have devoted myself unremittingly to natural history; in its pursuit I have travelled around the world. Facts in this science which before were inexplicable to me became one by one explained by this theory, and conviction has been thus gradually forced upon an unwilling convert" (92).

Moreover, Darwin's own conversion was not saltational but gradual. The evidence for this is meticulously examined by Sulloway (1982), in which he concludes that Darwin's "conversion to the theory of evolution did not spring full-blown as the result of his voyage, but emerged gradually in intimate cooperation with the numerous systematists who helped to correct many of his voyage misclassifications" (388). What is not examined by Sulloway is the evidence from Darwin's own recollections on the matter years later, repeated again and again, recollections for which there is no good reason for doubt or disbelief. For example, in a letter to Leonard Jenyns (October 12, 1844) he wrote "The general conclusion at which I have slowly been driven from a directly opposite conviction is that species are mutable & that allied species are co-descendants of common stocks" (Burkhardt and Smith 1987, 67). Around the time of the publication of the *Origin* Darwin emphasized this even more. For example, in a letter to James Dwight Dana (November 11, 1859) he wrote "It took me many long years before I wholly gave up the common view of the separate creation of each species" (Burkhardt and Smith 1991, 368). In a letter to John Stevens Henslow

(November 11) he calls it a "process" (370), while in a letter to Jenyns (November 13) he says "It took long years to convert me" (374). In a letter to John Gwyn Jeffreys (December 29) the language is identical: "It took me long years before I gave up the creation view of each species" (460). In a letter to Charles Bunbury written the next year (February 9, 1860) he stressed the point further, stating "I changed so slowly myself that I am indeed surprised at anyone becoming a convert" (Burkhardt *et al.* 1993, 76). Moreover, late in his life Darwin's memory of his conversion remained the same. For example, in a letter to Dr. Otto Zacharias (February 24, 1877) he wrote "in July, 1837, I opened a note-book to record any facts which might bear on the question. But I did not become convinced that species were mutable until, I think, two or three years had elapsed" (Darwin 1892, 175).[9]

Darwin's perception of the conversion of others was that it was also often gradual. In a letter to Cuthbert Collingwood (March 14, 1861), for example, he expresses "surprise" that he has "been successful in converting some few eminent Botanists, Zoologists, & Geologists," to which he adds "In several cases the conversion has been very slow & that is the only sort of conversion which I respect" (Burkhardt *et al.* 1994, 53). Again, in a letter to Alphonse de Candolle (January 14, 1863), he wrote "I remember well how many years it cost me to go round from old beliefs. It is encouraging to me to observe that everyone who has gone an inch with me; after a period goes a few more inches or even feet" (Burkhardt *et al.* 1999, 39–40). And in a letter to Dana (February 20, 1863) he repeats again that he values only gradual conversion: "Indeed I should not much value any sudden conversion; for I remember well how many long years I fought against my present belief" (155).

Perhaps a much better comparison for personal conversion in a scientific revolution, then, at least as indicated by the Darwin revolution, is not *religious* conversion but religious *apostasy*, the loss of faith, which tends to be gradual (cf. the collection of autobiographical essays in Babinski 1995, one of which is by yours truly). Often in such cases one does not simply give up one's former worldview; one also replaces it with another, a naturalistic one, a worldview that one thinks is both best and true. This is very different than the typical conversion from one religion to another. The latter is typically saltational, the former typically gradual. Of course in many cases the transition or conversion away from religious belief is for non-intellectual reasons: psychological, social, and in many cases even economic. But not all. Many cases are primarily for reasons of logic and evidence. Darwin's own loss of faith is a case in point (cf. Darwin 1876a, 85–96), and it was anything but saltational. Significantly, it seems to mirror his own conversion (and that of many others) to evolution, as being mainly a matter of the slow sifting of evidence and argument in spite of deeply-entrenched prior beliefs.

The case of religious conversion, on the other hand, where the conversion is from one religion to another, usually does not seem at all to fit this picture. The conversions of Pascal and Paul are but two of many cases in point. Here the reasons seem to be deeply psychological and emotional, not evidential and rational, which would explain why they tend to be saltational. In the case of Paul, for example, named Saul at the time, his conversion appears to have been the result of a buildup of guilt resulting from his zealous persecution of Christians, a buildup that finally peaked and erupted into a mental breakdown while on the road to Damascus, accompanied by hallucinations and temporary blindness. This is a gestalt switch *par excellence*, like a duck/rabbit or Necker cube, but it bears little if any similarity to the scientific conversions of Darwin, Hooker, Huxley, Gray, or Lyell, or of many others for that matter.

Indeed, what is of further interest here is that it would seem that it was not really the evidence *per se* that Darwin examined in his early years that converted him to evolution, but his genius in reworking and interpreting that evidence. On this view it would be his further genius, then, in developing his argument over the next 20 years, which played the main role in the conversion of other naturalists such as Hooker and Huxley. Sulloway (1982) points to "the fact that Darwin subsequently published this [early] evolutionary evidence in his *Journal of Researches* (1839) and in the *Zoology of the Voyage of the H.M.S. Beagle* (1841), fully two decades before the *Origin of Species* (1859). And yet not one naturalist appears to have been converted to a belief in evolution by these earlier works" (389). The real source for Darwin's conversion, then, claims Sulloway, was "Darwin himself," his "gifted individualism," and "not the evidence per se, that ultimately imposed the unorthodox interpretations that led him to embrace the theory of evolution" (389). Sulloway argues further that Darwin's genius did not stop there but continued in his later work, in which "he continued to do highly original work and to make important discoveries missed by his peers, in almost every branch of natural history to which he turned" (389). Interestingly, Sulloway's view (which I share), with respect to the 20 year period leading up to the *Origin*, seems to have been corroborated by one of Darwin's contemporaries, namely, H.C. Watson, who in a letter to Darwin (November 21, 1859) remarked "How could Sir C. Lyell, for instance, for thirty years read, write, & think, on the subject of species *& their succession*, & yet constantly look down the wrong road!" (Burkhardt and Smith 1991, 385).

What is interesting in all of this is the role of logic and evidence and individualism in concept change in a scientific revolution, something seriously ignored or at least underemphasized by Kuhn and by all those who follow him. But it is also often missed by those who try to reply to Kuhn by taking the route of the causal theory of reference. We have seen in the case of Beatty (1985) especially, the attempt to apply this theory to the case of Darwin on species. He took a modern philosophical theory on meaning and concept change and attempted to force it on the evidence, like trying to force a square peg into a round hole. And surprisingly everyone interested in the subject, including professional Darwin scholars who should have known better, followed him as if he had been discovering new facts. As remarkable as this all is, the fact remains that what Beatty did was revisionist history in the worst sense of the term. And it is that kind of history, though done in a very different way, that brings us to the next chapter.

Chapter 10

Darwin and the New Historiography

In this chapter I wish to take the analysis presented in the first eight chapters and use it to deal in a particular way with a small number of representatives of what may rightly be called in professional history of science "the new historiography." This is definitely not meant to include all or possibly even most professional historians of science flourishing in the past few decades, including those who focus on Darwin and topics Darwinian. It does not include, for example, those committed to what has been called the "Darwin industry," namely, the enormous amount of labor that has been put into publishing, and not just publishing but providing minute background details to, Darwin's unpublished manuscripts (Stauffer 1975), his articles (Barrett 1977), his notebooks (Barrett *et al.* 1987), his marginalia (Di Gregario and Gill 1990), and—the biggest job of all—his massive correspondence (which began with Burkhardt and Smith 1985 and which still, at the time of this writing, has the last seventeen years of Darwin's life to go). The Darwin industry also includes numerous interpretative essays largely based on these writings (e.g., Kohn 1985a), as well as the very fine biography of Darwin by Janet Browne (1995, 2002). Historians in this field, as David Kohn put it in his Introduction to Kohn (1985a), feel a "pressing need to place Darwin in the context of Victorian science," but they also share "the implicit premise . . . that Charles Darwin was a thinker of profound intellect and influence" (1).

Instead, it is the other end of the spectrum of modern history of science that I am mostly concerned with here, and it is what I mean by "the new historiography." It is usually called "externalist history" or "contextual history of science." In his biography of Huxley, Adrian Desmond (1997) uses the second of these phrases (xiv), telling us that his book is an example of "The newer approaches to science, emphasizing its class and social underpinnings" (xvi), that it "looks at evolution's use in order to understand the class, religious or political interests involved," that it is "a story of Class, Power, and Propaganda" (xiv). As such, it is, as he says near the end of his book, "a reaction to the old history of ideas, which *dis*placed the person, made him or her a disembodied ghost, a flash of transcendent genius" (617). Similarly, in their biography of Darwin, Desmond and Moore (1992) state that "Our *Darwin* sets out to be different—to pose the awkward questions, to portray the scientific expert as a product of his time; to depict a man grappling with immensities in a society undergoing reform" (xviii), or in other words "We want to understand how his theories and strategies were embedded in a reforming Whig society" (xix).

At its most extreme, the new historiography, in league with the "strong program" in the sociology of science, sees science as a mere social construction, its evidence and facts and theories no more objective or capturing reality than the stories of the Homeric gods, so that the distinction between externalist and internalist history is denied—denied because the internal development of scientific ideas (the ideational, intellectual, epistemic view of history) is itself denied (cf. Moore 1997, 290). In a sense, the new historiography is the exorcised ghost of behaviorism, with its emphasis on the environment, reincarnated into an entirely different discipline. But unlike behaviorism, which considered itself a genuine science, the new historiography is varyingly postmodern in its orientation, which denies that there is any real distinction between fact and fiction, truth and falsity, objectivity and subjectivity. Instead, looking to Nietzsche as their guide (which, incidentally, he would have despised), who proclaimed that the essence of all life is will to power, postmoderns dutifully denounce any claim to objective evidence and rationality as a disguise, as a mask hiding will to power, as Michel Foucault did against Noam Chomsky in his debate with him (Smith 1999, 181). Forget scientists testing theories against evidence, or their theories progressing toward truth; everything ultimately is a matter of will to power.[1]

Arguably there is much of this in Desmond's (1997) biography of Huxley, where in the very Introduction he tells us that "At the dawn of the twenty-first century 'reason' seems a precarious, value-laden yardstick, and one which has an infuriating habit of changing allegiance" (xv). Indeed he sees Huxley, in his effort to secularize and professionalize science and take it out of the hands of the clergy, as "establishing a rival evolutionary priest-hood" (xvi), such that his biography of Huxley is "the human story behind these sea changes" (xix). In all of this, Huxley the scientist is noticeably missing (cf. Lyons 1999). Given that Desmond's book is an attempt "To understand . . . the making of our modern Darwinian world" (xx), if we connect the dots he seems to be saying that our modern Darwinian world, including modern evolutionary biology, is merely a "sea change," subject to "changing allegiance." Kuhnian stuff at its most extreme (which Kuhn, in spite of the way he wrote, would always denounce as a misunderstanding of his view). Is this what is really behind Desmond and Moore's (1992) biography of Darwin? Michael Ruse (1993b) calls it "social constructivist" (229). In chapter 8 (note 15) we have already seen one example of the way in which their social constructionism operates. Later in the present chapter we shall see some further examples.

It only needs to be noted at this point how utterly alien social constructionism, along with its epistemological relativism, is to the modern scientific mind, for whom, as Dobzhansky (1973) approvingly put it, "nothing in biology makes sense except in the light of evolution." This quotation, not accidentally, is repeated and emphasized near the beginning of all three textbooks on evolution that I just happen to have at my home (Ridley 1993, 5; Freeman and Herron 1998, 51; Futuyma 1998, 3–4).

Most of the historians that I shall deal with in this chapter would, I strongly suspect, deny that they are part of what I call "the new historiography." Perhaps they all would. It does not matter. What we shall see is that, whether they rightly belong or not, they share in common—at least on the topic of Darwin on species—the following of a trend in their profession, namely, the desire to put scientists in general and Darwin in particular in his place, to embed him in his Victorian context, to keep him there, and to either diminish or deny his importance in what (typically by others) has often been called the most important (and still ongoing as I shall argue) scientific revolution of all time. It is all a matter of degree. Some professional historians do this more, some less. In this chapter I hope to try to counterbalance this trend, keeping of course my focus on Darwin and the nature of species.

Perhaps a good place to begin, and it will surprise many, is Peter Bowler's (1988) book, controversially titled *The Non-Darwinian Revolution*. To the naive browser at the local bookstore or library, this title might suggest that Darwin never started a revolution at all. Of course, creationists can put a lid on their hallelujahs, for once one reads the book one finds that what the title might suggest is not the reality at all. Bowler does indeed recognize that there was, or is, a Darwinian revolution, but he does not think it was started by Darwin's *Origin*. At the very heart of Darwinism, "the true essence of Darwinism" as he calls it (7), is not evolution but evolution by natural selection (in the form of adaptation), and it is natural selection, Bowler argues convincingly, that was rejected by the great majority of biologists from the time of the *Origin* right through to the forging of the Modern Synthesis, which began in the 1920s. (And even that, says Bowler, was not so much of a Darwinian revolution, since it in large part incorporated the genetics started by Mendel.) Thus for Bowler, the *Origin* did not start a Darwinian revolution; instead it merely served as a "catalyst" (5) for non-Darwinian evolutionary views, including mainly Lamarckism, orthogenesis, saltationism, and theistic evolution, all of them teleological to the core and all of them coming to an end in biology with the advent of the Modern Synthesis. Natural selection throughout this intervening period, when it was accepted, was typically seen as a minor mechanism in the pageantry of evolution. Hence the *non-Darwinian* revolution, the received view that the *Origin* began the Darwinian revolution being a "historical myth," a further example of "Whig history" (16).

There are two points I want to make about this, and they each have to do with the topic of Darwin on the nature of species. First, I am inclined to agree with Ernst Mayr (1985) that Darwin did not have one theory but five, and that together they are properly to be understood as what Bowler calls "the true essence of Darwinism." According to Mayr, these are evolution as such, common descent (that every living species is descended from an ancestral species), the multiplication of species (the branching conception of life), gradualism, and natural selection. Mayr argues that the first three of these theories were accepted by the great majority of naturalists within fifteen years of the publication of the *Origin* (1859) and that the remaining two theories became accepted as a result of the Modern Synthesis. Later, Mayr (1991) called the acceptance of the first three theories the "first Darwinian revolution" (12, 25, 107) and the acceptance of the latter two theories the "second Darwinian revolution" (89, 132). In other words, the Darwinian revolution is not a misnomer; it was neither a one-theory nor a one-episode affair.

That Darwin's *Origin* would start a revolution, and not be a mere catalyst for a non-Darwinian one, was perceived at the outset, and I don't think these testimonials should be disregarded. Watson, for example, upon reading the *Origin*, wrote to Darwin (November 21, 1859) "Your leading idea will assuredly become recognized as an established truth in science, i.e. 'natural selection'.—(It has the characteristics of all great natural truths, clarifying what was obscure, simplifying what was intricate, adding greatly to previous knowledge. You are the greatest Revolutionist in natural history of this century, if not of all centuries" (Burkhardt and Smith 1991, 385). Similarly, Wallace wrote in a letter to George Silk (September 1, 1860) "I have read it [the *Origin*] through five or six times, each time with increasing admiration. It will live as long as the 'Principia' of Newton. . . . Mr. Darwin has given the world a *new science*, and his name should, in my opinion, stand above that of every philosopher of ancient or modern times. The force of admiration can go no further!!!" (Burkhardt *et al.* 1993, 221 n. 1).

Darwin himself, of course, also recognized the revolutionary impact that his work would have, and he continually looked to posterity. In letters to Hooker, for example, he wrote "And what a science Natural History will be, when we are in our graves, when all the laws of change are thought one of the most important parts of Natural History" (July 30, 1856), "Whenever naturalists can look at species changing as certain, what a magnificent field will be open,—on all the laws of variation,—on the genealogy of living beings,—on their lines of migration &c &c" (July 13, 1858), "When we are dead and gone what a noble subject will be Geographical Distribution!" (December 31, 1858) (Burkhardt and Smith 1990, 194, Burkhardt and Smith 1991, 130, 231). Later in a letter to John Innes (December 28, 1860) he made the comparison to the Copernican revolution: "By far the greater part of the opposition is just the same as that made when the sun was first said to stand still & the world to go round" (Burkhardt *et al.* 1993, 540).[2] But it is in the *Origin* that we find Darwin's most specific remarks on the nature of the revolution that he envisioned would begin. There he says species will no longer be considered to have an essence (484–485), terms such as "community of type" will have a clear meaning (485), "A grand and almost untrodden field of inquiry will be opened up, on the causes and laws of variation, on correlation of growth, on the effects of use and disuse, on the direct action of external conditions, and so forth. The study of domestic productions will rise immensely in value" (486). He goes on to add that "Our classifications will come to be, as far as they can be made so, genealogies" (486). Psychology, he says, "will be based on a new foundation," and "Light will be thrown on the origin of man and his history" (488). Moreover, rather than evolution lowering the dignity of man, non-human animals will instead become "ennobled" (489). How utterly right was Darwin in every one of these, from antivivisection and animal cruelty laws to the discovery of the chromosomal basis of heredity to the discovery of DNA, to phylogenetic taxonomy, to paleoanthropology, to sociobiology and evolutionary psychology, through to animal psychology and ape language studies. What needs to be remembered is that Darwin, not in the *Origin* but in other works such as *The Descent of Man* (1871) and *The Expression of the Emotions in Man and Animals* (1872), himself opened many of these further fields of research, in particular, evolutionary ethics and evolutionary psychology.

Indeed I would go further than Mayr and suggest that there are really *three* Darwinian revolutions, the first and second having already occurred as Mayr argued (though with the exception of the species concept, since it is still in turmoil), the third being what we are currently in the midst of. Darwin himself recognized this third revolution, as the previous paragraph shows, but of course he could not possibly have envisioned the extent of that revolution. What I am referring to is the spilling over of Darwinian explanations outside the confines of biology as normally circumscribed. This third revolution has been ably discussed by Daniel Dennett in his book *Darwin's Dangerous Idea* (1995). It is arguably the most important (and still ongoing) scientific revolution of all time, which Dennett likens (approvingly) to a "universal acid" (63). Darwinism is a universal acid, for Dennett (and many others, e.g., Dewey 1910, 2, 19), not just because it ate through what were our traditional and most cherished beliefs in biology (e.g., creationism), but because it has leaked out into virtually all areas of life, including religion, ethics, politics, and psychology. This is the debate raging in virtually every university in the world today (cf. Stamos forthcoming).

But enough of this. What has any of the above to do with Darwin on the nature of species? Returning to the theme of Mayr's first Darwinian revolution, Darwin's *Origin* effected an enormous paradigm shift on the concept of species, changing it in roughly ten years or so from an essentially static, creationist concept to an evolutionary (or in some cases nominalistic) one. That was an enormous and truly revolutionary achievement. Granted, Darwin's own species concept was not understood (and who can blame his contemporaries for that?), moreover the new species concepts following the first revolution were teleological, as Bowler (1988, 5) rightly points out. But that should not be used to obscure the fact that Darwin effected a revolutionary change. For a start, he abolished essentialism for species. As long as species were conceived to be fixed types, there could be no evolution of species. Moreover, he radically raised the importance of variations (as well as varieties), in contradistinction to their previously low status. This change was accepted, in one way or another, throughout Bowler's non-Darwinian revolution. But just as important with all of this,

Darwin abolished the belief in what were thought to be two laws of species, belief (as Darwin well knew) that posed a major obstacle to the change from a creationist to an evolutionary view, namely, the belief in the law of limited variation and the law of reversion (cf. chapters 2 and 7).[3]

With regard to species and Mayr's second Darwinian revolution, Darwin, of course, did not effect a concept change, since (contrary to what Mayr and some others would like to believe) the concept of species in modern biology, if we look from the time of the Synthesis up to the present, was never settled. In fact it may even be thought to be more unsettled today than thirty years ago (Stamos 2003). That means, of course, that it is not over yet, and it still may be that two key features of Darwin's species concept—namely, his emphasis on the priority of the horizontal dimension for species reality as well as his advocacy of pattern over process, discussed in chapters 3 and 6 respectively—may eventually win the day. Only time will tell.

There is a second aspect of Bowler's book (1988) that I said I wanted to get to, but before I get to it we're still not quite finished with the first point. As I pointed out above, according to Bowler, Darwin eliminated teleology from his species concept. This is because variation, for Darwin, the very stuff that feeds natural selection, is essentially random with respect to the environment, and the evidence for this Darwin got from the breeders (29). But the *Origin* did not eliminate teleology from biology, not in the post-*Origin* period up to the Modern Synthesis. Instead, teleology re-emerged in various forms of non-Darwinian evolution, using the developmental model of embryogenesis, even in self-proclaimed Darwinists such as Haeckel (13, 83–90). As Bowler puts it, "the developmental model, in one form or another, continued to dominate late nineteenth-century thought. The materialist aspects of Darwin's theory which appeal to modern biologists were not typical of his own time; indeed, they were so radical that hardly anyone could accept them. . . . The antiteleological aspects of Darwin's thinking prized by modern biologists were evaded or subverted by the majority of his contemporaries" (5).

Interestingly, the claim that Darwin's species concept, at the core of his theory of evolution, was non-teleological has been denied by Robert Richards. According to Richards (1992), Darwin "conceived embryological evolution and species evolution as really two aspects of the same process: both kinds of evolution involved a gradual morphological change through a sequence of the same type patterns" (xiv). Again he says "I will indicate how, as I believe, embryological development became for Darwin a model of descent, infusing his conception precisely with those attributes, especially notions of progress, usually thought to characterize only non-Darwinian, romantic theories of evolution in the nineteenth century" (3). Noting that biologists today agree that three older proposals, common in the ninteenth century, should be dismissed—"that species evolution should be modeled on individual evolution, that embyogenesis recapitulates phylogenesis, and that evolution is progressive" (179–180)—Richards points out that Darwin in his early notebooks had all three. Dividing the "long gestation" of Darwin's evolutionism into three stages (which begins roughly when he returned from the *Beagle* voyage and continues through to 1859), Richards finds that in the first stage Darwin "considered species to be comparable to individuals," with a predetermined life span and the adapting mechanism of heritable effects of environmental agents. In the second stage, Darwin retained the adapting mechanism of heritable effects of environmental agents, gave up the predetermined life span, and took on branching evolution. There is no problem so far. It is the third stage that is troublesome. In this stage, says Richards, Darwin added the Lamarckian emphasis on habit and the inheritance of acquired characteristics (but without the Lamarckian component of conscious will), and he of course added as well (after reading Malthus) his own contribution of natural selection. According to Richards, "Though he regarded these 'Lamarckian' instruments as no longer central to the production of new species, he did retain them as various auxiliary mechanisms in the *Origin of Species*" (84). Again, still no problem. The problem begins when Richards claims that with natural selection, even in the *Origin*, Darwin maintained that "species evolution should be modeled on individual evolution" (179). As he puts it, Darwin considered evolution of the embryo and evolution of the species "as virtually the same process" (168). Again, he says Darwin used "embryological evolution as a model of species evolution" (169) and that "indeed, for Darwin embryological evolution became part of the causal matrix that produced species evolution" (169).

In all of this Richards (90–99, 104–108, 111–166) makes much of Darwin's apparent use of recapitulation in the *Origin* (that ontogeny recapitulates phylogeny). He claims to have "discovered that Darwin's theory, from its conception to its maturation, pulsed to the rhythms of that ever-fascinating principle" (xiv), so much so that Darwin's evolutionism contained "teleological factors, more than a whiff of which the history of recapitulation exudes" (176). Granted, as Richards recognizes, natural selection is a force operating on a species from without. Nevertheless, he sees Darwin being heavily influenced by, and thus retaining even in the *Origin*, the embryological model of evolution proffered by Continental embryologists. Although Darwin "did reject the hypothesis of an *intrinsic cause of necessary progress* buried in the interstices of organization," Darwin, says

Richards, simply externalized the cause so that natural selection, along with the subsidiary Lamarckian mechanisms, "would exert, as it were, an external pull, drawing most organisms to greater levels of complexity and perfection" (86).

This is a matter of ongoing debate, in particular between Richards and Michael Ruse (e.g., Ruse 1993a; Richards 2002; Ruse 2004; Richards 2004). One problem I find is with Richards' claim that the Continental embryologists were also species evolutionists. According to Richards, the recapitulation theories of Kielmeyer (19), Tiedemann (43–45), Treviranus (45–46), and Meckel (54)—but excluding Serres (167)—were wedded to a view of species evolution. None of the passages he cites in support of this interpretation, however, with the exclusion of Serres, are convincing. For example, he quotes Kielmeyer as saying "the force by which the series of species has been brought forth is one and the same in its nature and laws as that by which the different developmental stages [in embryogenesis] are produced." It is quite a stretch to read this as a belief in species evolution, especially given that the Great Chain of Being, believed in by almost everyone from the time of Locke to Darwin, was conceived to be an essentially static, nonhistorical series (Lovejoy 1936). Thus the "force" that Kielmeyer is referring to, for all that Richards' readers know, could simply be God, responsible equally for both the static series of species through time and also for the developmental stages in embryogenesis. The quotations from Tiedemann and Treviranus are hardly any better, and can easily be read as referring to species in the static Chain, given that the authors use words like "seems" (Tiedemann) and "appears" and "can be compared" (Treviranus). Meckel uses "probably," but we are not given the context so that we can determine what he means by "larger and smaller collections of organisms." Interestingly, Lyell (1832, 62–64) discussed the recapitulation theories of Tiedemann and Serres, but rejected them as providing any support for species evolution. Darwin, of course, would have read this, as he consumed Lyell (1832) while onboard the *Beagle*. He would also have noticed that Lyell does not ascribe to the two Continental morphologists a belief in species evolution. Instead he says their recapitulation findings "has appeared to some persons to afford a distant analogy, at least" to the theory of "progressive development" (62). For Lyell, instead, the recapitulation evidence only provides evidence of a "unity of plan" (64).

Another problem I find with Richards' thesis is his appeal to Darwin's use of recapitulation in the *Origin*. First of all, Darwin's acceptance of it is actually quite limited. Rather than say the development of the embryo of a higher mammal recapitulates the evolutionary stages of vertebrates from fish, to reptiles, to birds, to mammalian features, he simply tries to explain why this might appear so and he uses it as further evidence for evolution. For example, he says "the adult differs from its embryo, owing to variations supervening at a not early age, and being inherited at a corresponding age. This process, whilst it leaves the embryo almost unaltered, continually adds, in the course of successive generations, more and more difference to the adult" (338; cf. 439–450). This is simply an attempt to explain, from his own evolutionary point of view, something that many others had claimed to observe. Moreover, recapitulation for Darwin only "partially shows us" earlier stages (449), and he explicitly refrains from going as far as Agassiz in proclaiming recapitulation a law of nature[4]—which is not at all what we should expect him to do if Richards is right! Recapitulation was neverthe-less important for Darwin on two fronts, so that he hoped it would be proven a law (449). First, it provided further evidence for evolution, since "The point of structure, in which the embryos of widely different animals of the same class resemble each other, often have no direct relation to their conditions of existence" (439–440). In this sense it was like homologies, a point Darwin himself stressed (440). In either case the resemblances made no sense from a creationist point of view. Second, recapitulation, being a largely accepted part of embryology, would give further importance to em-bryology as an aid in classification, settling, for example, that a barnacle is really a crustacean (441). Once again, for Darwin, the importance is that "community in embryonic structure reveals community of descent" (449).

   In all of this it is quite a stretch to see teleology in Darwin's use of re-capitulation. To the extent that he accepted it, whatever that was, recapit-ulation was always for Darwin something entirely backward looking, not forward looking. As such, there is nothing teleological there. Embryology itself, on the other hand, was at the time of the Continental morphologists (with its etymological use of the word "evolution" in its embryology), as well as today (with its concept of the DNA program in development), en-tirely forward looking. Whether old or new, embryology is teleological. This difference, between Darwin's entirely backward-looking use of reca-pitulation and the forward-looking meaning of embryology, is enormous, and it seriously undermines any claim that for Darwin embryological evo-lution and species evolution "are virtually the same process."

   There is, of course, much more that should erode our confidence in Richards' thesis, and we have already seen much of it in this book. We have seen (i) that Darwin gave up the organism analogy for species in favor of the language analogy, (ii) that like languages he conceived of species as pri-marily horizontal entities, (iii) that his species concept was a pattern species concept, not a process species concept, and as such could have no teleology in it, (iv) that he avoided the word "evolution," preferring "descent with modification" instead, and refused to call his theory "developmental," (v) that in the *Origin* he called variation under domestication "our best and safest clue" (4), (vi) that he repeatedly called species "plastic" (12, 31, 80, 132), (vii) that he maintained throughout that their plasticity followed from his view that variation, without which natural selection can do nothing (82), is random with respect to the environment (30, 46, 61, 81, etc.), and (viii) that in spite of his language in the *Origin* he did not think of natural selection as a real force or power.[5] What more could one possibly want for a non-teleological species concept and a non-teleological process of evolution?

Richards, however, seems consumed with selectivity of evidence. He recognizes that the received view gets support from Darwin himself, in his repeated claim "against Lamarck's idea of an 'innate tendency toward progressive development'" (85). In his footnote, Richards (n. 45) cites as his source an 1872 letter from Darwin to Alpheus Hyatt, and he says a similar remark is to be found in Darwin's *Sketch of 1842* (Darwin 1909, 47). And of course there is much more in this vein (e.g., Darwin 1859, 412, 351, 1863a, 79, 1876a, 87; Burkhardt *et al.* 1993, 577–578). But the real point is that, all things considered, Richards has provided a paltry effort at examining contrary evidence. Instead, in his book he continually presses on with his thesis, trying to shove his square peg into a round hole.

It seems indisputable that the force of ideology, not evidence, is here at the core. Richards says the received view on Darwin's non-progressionism is itself laced with "the darker byway of ideology" (xv), that ideology being the Modern Synthesis, which eschews recapitulation and progress (174–180). Richards ever so briefly hints that his own reinterpretation of Darwin also has behind it a motivating ideology, when he writes "My own narrative undoubtedly depends on some considerations that bend history in ways that would appear distorting to others" (174). But anything more than this is met with silence. I suggest that the ideology that drives Richards is a common one among modern historians, the one we find in Desmond and Moore and in other authors examined in this chapter, namely, *to keep Darwin in his place*, to not allow him to be a man ahead, let alone well ahead, of his time. This is the rallying cry of the new historiography, with respect to scientists in general and to Darwin in particular, and with which I am here in dispute. And indeed at the very end of his book Richards seems to briefly show his hand, when he writes "Darwin was indeed the architect of the theory that has been reconstructed as neo-Darwinism. But the architect was our ancestor, who dwelt happily enough in the nineteenth century" (180).

As if to further appease adherents of the received view, Richards adds at the beginning of his book that "The Darwin that emerges from this study will appear decidedly more venerable than the rejuvenated evolutionist who has been injected by some historians with the monkey glands of a modern scientific ideology" (3–4). But it is hard to see how Darwin becomes more venerable from Richards' study. Rather than more, it would seem less. And that indeed seems the whole point to the new historiography.

I said earlier that there is a second point to Bowler's book that I would get to later on, and here is a good place to deal with it. Noting that Darwin had not only read Malthus but also Adam Smith and other political economists, Bowler (1988) writes

> The concept of divergence through specialization reflects the economic advantages supposed to accrue from the division of labor. Even the Darwinian concept of species seems to reflect the individualist model of society. The orthodox typological view of species can be compared with a totalitarian political philosophy; individual organisms must conform to the specific type just as individual human beings are supposed to submerge themselves in the higher reality of the state. To some extent, Darwin saw the species as a population of varying individuals with no fixed type, just as *laissez-faire* economics sees society as composed of individuals with divergent interests. [36]

This is exactly the sort of thinking that one would expect from the new historiography, and it is particularly interesting since it focuses on the choice of species concept, as something culturally determined. It is not entirely clear whether or to what extent Bowler himself subscribes to the view that he expresses here. The passage appears in the context of Bowler's discussion on the debate between the Darwin industry (which Bowler says basically adheres to the myth of the Darwinian revolution) and the sociologists of science. Bowler deserves the benefit of the doubt, however, since immediately following the above passage he states "The possibility that Darwinism reflects certain aspects of Victorian society is accepted by the majority of modern scholars, but this does not commit them to the view that natural selection is nothing more than a projection of the capitalist ethic onto nature" (36).

Contrary to what Bowler suggests at the time of his writing, it would seem that epistemological relativism, in the form of sociocultural relativism, has become much more common in professional history of science. Indeed Bowler's statement above was written just before the books by Desmond (1989) and Desmond and Moore (1992), both of which received book awards and have been hailed by an increasing number of professional historians as paradigms of history of science writing (cf. Mauskopf 1992, 279; Shortland and Yeo 1996, x)—and one can only imagine the flock of graduate students following. Times have indeed changed since 1988. Hence David Hull (2000) writes of the "sociocultural turn" (72) in history of science, with its "epistemological relativism" (75) and its "postmodernists" (80), and Mary Winsor (2001) makes a powerful appeal for the "emancipation" of professional history of science from its self-imposed "Taboo Problem," a taboo that amounts to "blacking out the very center of the topic we claim to study" (240), the center being the "progressive direction of scientific change" (241)—by which she clearly means epistemological progress.

At any rate, in the case of Darwin the new historiography tends to make much of the supplanting of the old aristocracy by the rising middle class of the Industrial Revolution, of which Darwin's family, on both sides, was part of. It also tends to make much of the influence of the backward Victorian attitude toward sex. I shall return to the latter thesis below. Beginning with the former, one could easily relate Darwin's view of diverging species, of species evolution driven not only by natural selection but also by ecological divergence, with the emphasis on the division of labor by the capitalist economists of Darwin's time. In the *Origin* Darwin himself hints at a relation between the two, stating that "The advantage of diversification in the inhabitants of the same region is, in fact, the same as that of the physiological division of labour in the organs of the same individual body—a subject so well elucidated by Milne Edwards" (115–116). In *Natural Selection* (Stauffer 1975, 233) the analogy is likewise to physiology, but elsewhere and earlier in his notes Darwin explicitly drew the analogy to economics (cf. Ospovat 1981, 181). Desmond and Moore (1992) make much of this. The problem for Darwin was to explain the branching, treelike diversification of nature, reflected in the static hierarchical classifications of his fellow naturalists. The answer for Darwin, claim Desmond and Moore, came not from nature itself but from the political economy that dominated his society as well as his own personal life. Thus not only do they describe Darwin's mechanism of natural selection as "his Malthusian, capitalist, competitive mechanism" (413), but we're told that "Just as his Malthusian insight had come from population theory, so his mechanism for creating diversity looked like a blueprint for industrial progress. Darwin was a heavy investor in industry. His Wedgwood cousins were among the pioneers of factory organization. They created the production-line mentality with a marked division of labour among the work force" (420). Again, for Darwin "The creation of wealth and the production of species obeyed similar laws. Division of labour was nature's way as well as man's" (420–421).

Again, Darwin used "the zoologist Milne-Edwards's use of the term 'division of labour' rather than an economist's'" because "A political taint would have made natural selection too much of a target" (421). We are told furthermore that Darwin had alternative models to choose from, such as that of the economist Jean Sismondi (his wife Emma's uncle), according to whom the division of labor was unjust and those who worked the land should own it. But Darwin rejected this and other models, we're told, because "he was himself an absentee landlord in Lincolnshire" (421). We find that Darwin "put his money where his mouth was," that "He spent tens of thousands of pounds on railway companies, and twenty years of his life revealing the competitive, specialized, and labour-intensive aspect of Nature's 'workshops,'" that "He was placing Nature on industry's side" (421).

In all of this there is nothing about evidence but only hidden motive. A theory of motive, however, is an empirical thesis, and empirical theses don't become facts simply by being claimed. What is required is supporting evidence. There are certainly interesting correlations between the socioeconomic conditions in Darwin's time and Darwin's theory of evolution by natural selection and divergence. But correlation is not causation, and any argument from correlation to causation needs something more—something much more, as students of fallacies and medical researchers know only too well. But Desmond and Moore don't seem much interested in weighing evidence pro and con. Correlation seems sufficient for them. Nor do they seem interested in the principle of charity, giving Darwin the benefit of the doubt until there is strong enough evidence against him. Instead, they have an agenda, and they spare no effort in pushing it. For example, barely (448) do they mention Darwin's enormous work on small and large genera, in which he argued, from tables of classifications by naturalists and related information supplied by Watson, Hooker, and Gray, that species in large and wide-ranging genera have more varieties than species in small genera, important corroboration for his principle of divergence (Stauffer 1975, 134–164; Darwin 1859, 53–59). Perhaps the numerical analysis he accumulated was not after all good from the viewpoint of modern statistics (Parshall 1982). Of greater significance is what Kohn (1985b) has argued, that Darwin's principle of divergence was of the nature of an internal dialogue motivated by explanatory unification, that the theory came first and the attempt to test it came shortly after. From his biogeographical studies Darwin came to realize that natural selection, even when combined with geographic isolation, seemed insufficient to explain hierarchical classification (something recognized by all, and that Darwin temporalized as treelike). Natural selection therefore needed something more, which he found in the concept of "places in the economy of nature," sympatric speciation meaning for him that selection favors diversification within a species. What is surely significant is that Darwin, as Kohn (1985b, 256) has shown, developed his principle of divergence *first*, in November of 1854, and then only *later*, much later in September of 1856, connected it to economics and the division of labor (cf. also Ospovat 1981, 181). As Kohn puts it, "To mistake the labeling for the conception would, I believe, be a misinterpretation of Darwin's developmental process" (256).[6]

It is still possible that in developing and emphasizing his principle of divergence Darwin was influenced by the political economy of his time, and perhaps the emphasis on *laissez-faire* economics influenced his species concept with its varieties within a species, along with the importance of the individual variants. Perhaps also Darwin's (and Wallace's) "discovery" of natural selection was not really a discovery but a product of the same political economy, projected onto nature, and that Darwin's continued allegiance to natural selection is to be explained by this environment (cf. Radick 2003 and Bowler 1988, ch. 2 as a counterbalance). But to "pull focus"—to use Moore's (1997, 296) cinematic metaphor—like that, to shift focus from the individual in the foreground to the environment in the background, is to miss so much more that was going on, such as Darwin's observation of variation in barnacles, his consilient argument for evolution by natural selection (which included especially the evidence from breeders and biogeography), as well as his argument against the so-called laws of limited variation and reversion. Kohn (1985b) offers the following reconstruction concerning Darwin's principle of divergence: "As he [Darwin] comes to see the species of local genera as the primary locus of divergence, he comes to see small locales with no chance of geographic isolation as the primary cites of speciation. Appreciating full well the swamping effect of crossing, he is forced to invoke vigorous selection as the only effective countervailing force. But more important, this line of thinking leads Darwin to look for the local, hence ecological, conditions that favor vigorous selection" (255). To ignore the role of problem, evidence, internal dialogue, and testing in Darwin's theorizing is simply poor scholarship.

Rather than there simply being a social constructionist influence from the political economy of Darwin's day (to which he seemed to subscribe) to Darwin's theory or principle of divergence, it may well be that Darwin keyed into an important process that actually exists in multiple domains. We have seen in chapter 3 that Darwin argued that natural selection is not domain specific, that it also functions in language evolution. This is an insight embraced by not a few linguists today (e.g., Lass 1997, 376–381). Many today would also argue that Darwin recognized real similarities across two other domains, namely, economics and ecology, including the adaptive response of diversification and specialization. In fact a whole journal now exists for the single purpose of exploring these and other connections between biology and economics, appropriately titled the *Journal of Bioeconomics*, founded in 1999 by Michael Ghiselin and Janet Landa. To dismiss all of this as mere social construction or will to power is the easiest thing to do, as long as one is not willing to take evidence seriously. The real skill is to make inferences to the best explanation based on all the evidence available, but many, influenced by certain intellectual fashions, do not seem willing to do this.

Whether multiple domains or domain specific, it is surely interesting, and a further point in Darwin's favor, that in recent years, in a longstanding economic climate that embraces competition but disdains *laissez-faire* economics, biologists have more and more been reassessing their previous rejection of sympatric speciation (e.g., Ghiselin 1997, 159), so much so that some biologists now claim that of *all* speciation models "the mechanism of sympatric speciation proposed by Darwin . . . is the most plausible one" (Kondrashov *et al.* 1998, 97).

Another area the new historiography likes to make much of is sex, in the case of Darwin and his contemporaries the Victorian attitude toward sex. Desmond and Moore (1992) relate Darwin's theory of sexual selection to his family life, in that "The Darwins fit the picture perfectly. The *Descent* was essentially their story. Natural and sexual selection had made and maimed them. Charles had strutted like 'a peacock admiring his tail' courting Emma. Coy and impressionable, she had selected him, admiring his 'courage, perseverance, and determined energy' after a voyage around the world" (580). *She* had selected him, never mind that *he* had seriously weighed the pros and cons of marriage, as shown in his notes on marriage of April and July 1838 (Burkhardt and Smith 1986, 443–445).

Here, as elsewhere, one can easily get carried away with social constructionism. Gordon McOuat (2000), in his review of volume 10 of Darwin's *Correspondence* (Burkhardt *et al.* 1997), hints that the emphasis by Huxley and others on the sterility criterion for species had a Victorian ring to it. As McOuat puts it, for Huxley "Sexual sterility marks a true separation between species, and if natural selection could produce sterility, it could provide the origin of new species. At the very ground of the division of natural kinds in biology was a sexual demarcation. There was something naughty, forbidden here. In one way it rubbed against Victorian sensibilities" (191). My God should Huxley now be thought more of a good Victorian? His species were, after all, good Victorian species. And what of Darwin? Was he a decadent? Unlike Huxley, his species were bad Victorian species, some-times with sterility within them,[7] but worse—much worse—sometimes with fertility between them. Surely there is sexual deviance afoot here, evidence of a desire for promiscuity.[8] Perhaps even bestiality! And wasn't all of this really what was behind the veil covering the Victorian attitude toward sex? Didn't Freud teach us that? Even more, Darwin's rejection of "higher" and "lower" in nature (e.g., Notebook B 74, 1859, 441, Burkhardt and Smith 1991, 228) may well be read as a symptom, and along with it as his own pro-jection onto nature, of an underlying decay in Victorian society with regard to its concept of "place" (i.e., the Victorian caste system with its layered aristocracy at the top and the homeless poor on the bottom).

On the other hand, as we have seen in chapter 3, Darwin himself did prefer a horizontal to a vertical species concept. This cannot be ignored, and it should make one rethink one's history. Correlation is causation, after all. Somehow, someway, Darwin must be seen as "a cork bobbing on the sur-face of the society of his day," as Ruse (1993b, 229) characterized the new historiography. Thus it may well be that Darwin was, in a sense, more of a good Victorian after all, and by that I mean the overtly Victorian sense, not the Freudian. Darwin almost certainly was influenced, probably sub-consciously, by the Victorian insistence on the missionary position when he conceived of species as primarily horizontal. And as I have shown in chap-ter 3, he maintained that view throughout. His species were indeed, there-fore, in a very different sense than before, good Victorian species (this was his *real* torment). All of this, interestingly, can be traced back to Darwin's *Beagle* voyage, during which he got to observe much of the work of the Christian missionaries, and we know from Desmond and Moore (1992, 173–176) that contrary to the critics back home he approved of their moral effect. It is therefore only reasonable to suppose that Darwin knew very well of the attempts by the missionaries to impose the horizontal position on the poor savage aboriginals for when they conceive their children (hence what we now call "the missionary position"). And probably this influence on Darwin's species concept was further reinforced by personal experiences in his life, some of them not even overtly sexual, again such as his *Beagle* voyage in which during seasickness "only a 'horizontal position' brought relief" (Desmond and Moore 1992, 115).

McOuat (2000) goes on to say "Freudians could have a field day with Darwin's envious astonishment over the size of the barnacle's penis," and that in the correspondence "The married Darwin flirts; his sisters tittle" (191). Did Darwin have an inferiority complex over his penis? Was this common in Victorian society? Did he try to make up for his feelings of inadequacy by desiring sex outside the marriage? Was promiscuity seething beneath the Victorian exterior? I do not say that McOuat is guilty of such nonsense (he is not), but his remarks nevertheless invite the kind of garbage that social constructionists seem to thrive on. It is tabloid historiography, with all the irresponsibility and sensationalism that goes with the tabloid journalism found in the neighborhood grocery store. Book awards, moreover, don't change this fact one bit.

Leaving my parody of the new historiography aside, a further area of cultural influence that the new historiography likes to focus on is the scientific culture of the time. Sometimes issues of scientific culture cannot be clearly separated from politics and culture at large. This is the theme of Harriet Ritvo's (1997) article on zoological nomenclature, where, following Foucault's work (1971) on the eighteenth century, she finds the Victorian debates driven mainly by the desire for dominion and control, whether geopolitical or personal, including also patriotism and vanity. There is much in her analysis that deserves merit and serious consideration, and yet, in a way that has become all too typical among professional historians of science, she goes too far. Confining ourselves to the case of Darwin, granted, Darwin was involved in and supported the rules drafted by Strickland's 1842 Committee. But with regard to an earlier draft he demurred (seconded by Leonard Jenyns) on the side of foreigners where it stated that the British Association for the Advancement of Science should properly be considered "the Parliament of Science" (Burkhardt and Smith 1986, 311). Moreover, he pointed out in a letter to Jenyns (May–September 1842) that "Stricklands laws will I think be useful in checking the egoism of some authors" (317). Darwin himself, of course, had become impressed with nomenclatural problems during his stay in London following his *Beagle* voyage and the cataloging of the thousands of specimens he had brought back with him. Years later while working on barnacles, in which he had inherited a nomenclatural mess from his fellow naturalists, the need for rules and consistency became even more important. But in none of Darwin's practices do we find political or egoistic concerns. He respected the priority of foreign authors (cf. Darwin 1854b, 2), and even when he felt the need to go against his fellow naturalists and call a form a "species" he did not use his own name (e.g., Darwin 1854b, 250). Nor in naming, for example, primroses and cowslips two species instead of one was he trying to do something like raise the dignity of the common man (who named them two species instead of one, contrary to scientific practice), or any other such nonsense that the new historiography might come up with. He kept the names the same. Instead, whenever he tried to correct the species ascriptions of his fellow naturalists he did so purely for what he believed to be the soundest of scientific principles, namely, common descent partitioned by adaptations produced by the law of natural selection.

This raises a further issue. Sometimes the influence of scientific culture is mixed very little if at all with the wider influences of politics and culture at large. We have seen some of this already earlier in this chapter in Richards' (1992) attempt to make out of Darwin a *Naturphilosoph*. A much more plausible approach is to look at the scientific culture of Darwin's day with its emphasis on explanatory laws of nature, along with its emphasis on natural theology by way of the many beautiful adaptations to be found in the living world. There is surely an influence here on Darwin's species concept—arguably more than any other area of influence—with its emphasis on adaptations and their explanation by natural law. Not only was the discovery of laws of nature highly prized in Darwin's day, to the point of calling phenomena laws that were not really laws, but laws also played a central role, as we have seen, both in scientific explanation and scientific classification. So great was this influence that Darwin called community of descent a *vera causa*, even though this conflicted with his intuition that the extinction of a species is not necessarily forever.

However great the influence of this culture was on Darwin, we nevertheless still have to be very careful. For a start, adaptations were then as now an extraordinary fact of nature, and they cry out for explanation. Darwin of course knew this, but his importance lies in the fact that he was the first person in history to recognize that their explanation must be by natural selection.[9] Modern biologists have become absolutely certain that Darwin was right here, and for this (and much more) they rightly honor him (e.g., Dawkins 1986, ix, 6, 287).

In spite of Darwin's importance here, however, which as we have seen in this book is the key to understanding his species concept, some historians have tried to minimize Darwin's importance by reevaluating the developments in taxonomy prior to the birth of Darwin's crowning achievement. Thus Mary Winsor (2003), for example, who for reasons given earlier I greatly hesitate to include among the ranks of the new historiography, has attempted to rewrite history by arguing that pre-Darwinian taxonomists, contrary to the received view, were not in fact essentialists. There is nothing new in this claim. It has been made, for example, by McOuat (1996), as we have seen in chapter 8. What is new is Winsor's claim that pre-Darwinian taxonomists were clusterists (my term, not hers). Never mind what they stated in their definitions, she argues, it's what they did that matters, and their practice clearly indicates that they were not essentialists. Strict essentialism classifies according to one or more conditions individually necessary and jointly sufficient for membership in the class. Clusterism, on the other hand, involves a set of membership conditions, but no one condition is either necessary or sufficient for membership in the class, so that membership is a matter of degree. Clusterism, according to Winsor, is how we should characterize pre-Darwinian taxonomists. What is more, following the modern philosopher Richard Boyd (1999), who argues that species should be conceived as homeostatic cluster classes, Winsor argues that because clusterism does not in itself pose a barrier for evolutionism we should therefore look at pre-Darwinian taxonomy as paving the way for the Darwinian revolution, indeed as "constituting the foundation of Darwinism" (398).

The problem with her argument, at bottom, is not that Boyd and others might be wrong, that homeostasis is a problematic concept and clusterism is arguably not compatible with evolutionism (as I have argued elsewhere; cf. Stamos 2003, 224–230, 123–143). Instead, the fundamental problem with her argument is that when dealing with pre-Darwinian taxonomists she does not discriminate between species and higher taxa. In fact, all of her evidence for clusterism in the practices of pre-Darwinian taxonomists comes from higher taxa. In my public debate with Winsor and in my subsequent reply paper (Stamos 2005, 81–83), I provided further evidence to support her in this area. For example, in a letter to Darwin (April 26, 1844) George Waterhouse, a well-respected entomologist and member of Strickland's 1842 Committee, wrote "The term 'typical species' is used by Zoologists in two senses—it either refers to that species which possesses in the highest degree of development *some* of the characters which distinguish the group ['I mean *any* assemblage of species'] to which it belongs from other groups; or, it has reference to that species which is supposed to exhibit, in the best balanced condition, the greatest number of characters common to the species forming the group of which it is a member" (Burkhardt and Smith 1987, 30). Either way we have clusterism here, and the evidence is overwhelming for such clusterism at the level of genera and higher taxa. It was indeed the dominant view.

What Winsor has missed, however—and it is why we cannot afford to not pay attention to what these naturalists say, including their definitions, if we are to put labels on them—is that, again as I argued in the debate and in my paper, the great majority of pre-Darwinian taxonomists did not believe in the reality of higher taxa from genera up. Buffon, Jussieu, Prichard, Bentham, Lindley, Watson, and Hooker—representative examples with enormous influence—were all higher taxa nominalists. Typical and uncontroversial here is what Agassiz (1857) wrote, that in spite of numerous different systems of classification "There is only one point in these innumerable systems on which all seem to meet, namely, the existence in nature of distinct species persisting with all their peculiarities . . . . Beyond species, however, this confidence in the existence of the divisions . . . diminishes greatly. With respect to genera, we find already the number of naturalists who accept them as natural divisions much smaller; few of them have expressed a belief that genera have as distinct an existence in nature as species. As to families, orders, classes, or any kind of higher divisions, they seem to be universally considered as convenient devices" (4–5). For most of these naturalists, then, they grouped species into higher categories simply out of practical necessity. To call them clusterists is highly misleading. Methodologically they were clusterists, but ontologically they were nominalists. For the great majority of them, only species were real.

It is therefore an enormous mistake to look at these pre-Darwinian naturalists and taxonomists, including Darwin's contemporaries, and paint them with the same brush when it came to species. Not every received view is mistaken, although professional historians of science often seem bent on disproving one received view after another. In the case of pre-Darwinian naturalists and taxonomists, not only is the received view correct in that most of them were species realists, but the majority of them were also species essentialists. Again the evidence for the received view here is overwhelming. For a start, as I argue more fully in my reply paper (Stamos 2005), most of these naturalists and taxonomists believed in primordial descent, with each species beginning with a single individual or pair. Second, most of them were splitters, not lumpers, not allowing for varieties and making the varieties of lumpers into species. Third, most of them believed in reversion as a law, which allowed for some variation in characters but those characters would revert back to the primordial type following the removal of the perturbing conditions (typically they thought this law relevant only for domesticated species). Fourth and finally, most of them believed in an underlying causal essence, passed on through generation, which accounted for reversion and limited variation. We have seen some of this in chapter 7, in Linnaeus' "unity in generation" where variation is in the "outside shell," in Buffon's *moule intérieur*, in Jussieu's "any individual whatever is the true image of the whole species," in Agassiz's essences "forever fixed" in the mind of God, and in Dana's "specific law of force, alike in all." Moreover, as we have seen in chapter 1, what went hand in hand with this essentialism was what Wollaston called the "axiom" of independent creation.

With the concept of species at its core, then, what Darwin's revolution signifies is an enormous departure from the paradigm that existed before. Pre-Darwinian taxonomists and naturalists, contrary to Winsor, did not lay the foundation for Darwinism. Instead, they posed an enormous barrier. To think and argue otherwise is to play into the hands of the new historiography. It is, yet once again, to try to minimize Darwin's importance, to see him as a mere extension of his culture.[10] It is to follow a facile, myopic, and ultimately pernicious trend.[11]

If we really want to see science as a social construction, we can do no better than look at the Lysenko affair that occurred in Communist Russia beginning in the mid-1930s (cf. Harman 2003). This episode illustrates as perfectly as can be what happens when science really is a social construction, when it is based primarily on culture and ideology rather than on nature. Previously among the leaders in the world, Russian biology under Lysenko—with its forced rejection of Mendelian genetics (including the imprisonment and death of many geneticists), and also of Darwinian selection on random variation, both of which were viewed as contrary to Marxist dialectics, unlike the naive Lamarckism that Lysenko made mandatory—came to a grinding halt, with enormous negative consequences for its agriculture. Only in the mid-1960s, when Lysenkoism was officially dismissed by the Soviet state, did Russian biology and agriculture begin to resume where they had left off, trying to play catch-up ever since.

The Darwinian revolution, on the other hand, with its five theories at its core, has proved the exact opposite. It has opened up one fruitful research program after another, leading to spectacular successes in knowledge and understanding, so much so that biologists today routinely agree with Dobzhansky that "nothing in biology makes sense except in the light of evolution." Darwin was a Victorian, to be sure. His was a time very different than our own in many ways, of course. But in so much of his thinking he belongs more in our day than his own. Biologists have recognized this time and again, and not without good reason. Natural selection, for example, has been amply documented in the wild (Endler 1986), it has proved essential in understanding phenomena such as the evolution of bacterial resistance to antibiotics (Sonea and Panisset 1983, 86–87) and insect resistance to pesticides (Ridley 1993, 101–106), and it is the key to understanding immune systems and the evolution of viruses such as HIV (Freeman and Herron 1998, 3–25). In fact biologists, and to a lesser extent philosophers, continue to find helpful insights and inspiration in Darwin's writings. Sexual selection is a good example, since the process itself, as well as Darwin's insights into it, have come to be appreciated by mainstream biology only in the past three decades (Andersson 1994). Sympatric speciation is an even more recent example (Kondrashov et al. 1998). While outside of biology proper, Darwin's principle of divergence has come to be seen as a useful explanation of divergence in personality types among siblings (Sulloway 1996, 84–85). Indeed evolutionary psychology, begun by Darwin, finds that it owes much to Darwin (Crawford and Krebs 1998), as does evolutionary ethics (Thompson 1995).

Darwin still has much to teach us in many areas, in spite of his mistakes and shortcomings here and there. With mass extinction #6 in full swing as the twenty-first century moves forward, now more than ever we need to learn and take into full consideration what he has to say about the nature of species. In his own time it would have fallen on deaf ears. But to ignore him today, or to diminish his importance as the mere product of his time, is simply arrogant folly. I do not know whether Darwin's insights into the nature of species, as I have brought them to light in this book, will ever experience any of the good fortune that so many of his other ideas have. (Whiggish history this book is not, although my hope is that it someday will be read as if it were.) But whatever the future holds, or should hold, surely the time has come for Darwin to get a fair hearing on this head.

Notes

CHAPTER 1

1. Throughout the present book all references to the *Origin* are to the 1859 edition, unless otherwise indicated.

2. This was indeed a common view of genera in Darwin's day and before, even though it was not shared by Linnaeus. For example, Bentham, who Darwin explicitly refers to on this topic on p. 419, had long maintained (Bentham 1836) that "A genus . . . has seldom any real existence in nature as a positively determined group, and must rather be considered as a mere contrivance for assisting us in comparing and studying the enormous multitude of species, which, without arrangement, our minds could not embrace" (xlvii; cf. Bentham 1858; cf. also Stevens 1994, 99–109, for references on higher taxa nominalism in Bentham and in other naturalists). In like manner Watson, in a letter to Darwin (March 23, 1858), confided that "I look upon the orders & genera of plants as purely conventional arrangements, not natural groups,—that is, not groups *in nature*, but only as groups in books & herbaria" (Burkhardt and Smith 1991, 54). Similarly Hooker, in a letter to Darwin (February 25, 1858), stated that "Genera in short are almost purely artificial as established in Botany" (Burkhardt and Smith 1991, 35). That genera and higher tax nominalism was the general view, cf. also Watson (1845a, 142) and Agassiz (1857, 4–5), as well as Stamos (2005).

3. It is sometimes claimed that this was in fact not a common feature of species concepts in Darwin's day (e.g., McOuat 1996, 511, 515). And yet it is easy to find contemporary sources who agree with Darwin on this point. For example, Watson (1845a) states that "the very definition of the term 'species,' as usually given, involves an assumption of non-transition" (147). He would later repeat this in a letter to Darwin (May 10, 1860), in which he wrote "Until a faith in certain impassable barrier between existent species becomes thoroughly shaken, naturalists will resist your views, & hail difficulties as if conclusive arguments on the contra side. Differently as these unseen barriers are traced or placed, they are believed in about as strongly by almost all" (Burkhardt *et al.* 1993, 203). More specifically Wollaston (1860), in a critical review of Darwin's *Origin*, states that "The opinion among naturalists that species were independently created, and have not been transmitted one from the other, has been hitherto so general that we might almost call it an axiom" (133).

4. Interestingly, Darwin seems to have got more from Watson than just the above strategy. For in his correspondence with Darwin, Watson, in spite of what he says about the variability in species concepts between his fellow naturalists, seems still to have thought that species are in a sense real entities in nature. For example, in a letter to Darwin (October 11, 1855) he wrote "I look upon the words 'Orders, genera, species (of books), & varieties,' only as terms to indicate passably well the grades of resemblance between objects" (Burkhardt and Smith 1989, 479). That Watson includes only book species is highly significant. Elsewhere he made explicit his distinction between *book* species, which he considered arbitrary, and *natural* species, species found in nature (Watson 1843, 618). Granted, a little later (Watson 1845b), as we shall see below, after focusing on the messy situation presented by primroses and cowslips, and how they lend themselves to evolutionism, he came close to drawing a nominalistic conclusion about species (219). However, he never went all the way, and in later writings, in spite of his evolutionism, he seems clearly to have thought that there are real species in nature, moreover that their reality is only or primarily horizontal in nature, at a given time slice. For example, in a letter to Darwin (November 8, 1855) he wrote "I must confess a pretty strong bias towards the view, that species are *not* immutably distinct;—altho' in our time-narrowed observation of the individuals they seem to be so" (Burkhardt and Smith 1989, 499). Similarly, in a later letter to Darwin (December 20, 1857), he wrote "In writing the final volume of my Cybele Britannica, I find myself unable to carry out the ideas or inquiries originally intended. And why?—Mainly, because the limits of species are so uncertain in nature,—so dissimilar in books. . . . This leads me to devote many pages to my own notions about species & classifications,—rather irrelevant in a book on local botany;—and perhaps somewhat limping over that same ground which will be better trod by yourself. I cannot find the proof of species being definite & immutable, whatever they may seem to be at any one time & spot" (Burkhardt and Smith 1990, 511). Cf. Watson (1859, 31–34), where he distinguishes between true species in nature, which he defines from an evolutionary point of view (implicitly horizontal), and what he calls "*false* species," for which he provides four tests. Watson's repeated emphasis on the general horizontal reality of species will prove especially significant when we look at Darwin's own view in Chapter 3.

5. In spite of Lewes' confusion between species as a category and species as a taxon (cf. Chapter 2), I find Lewes' position extremely interesting for the simple reason that after the above passage he immediately proceeds to provide an ontology for species based on similarity relations, and one that happens to be close to what I shall argue in later chapters was in fact Darwin's view. According to Lewes, "Nature produces individuals; these individuals resemble each other in varying degrees; according to their resemblances we group them together as classes, orders, genera, and species; but these terms only express the *relations of resemblance*, they do not indicate the existence of such *things* as classes, orders, genera, or species" (443). Later I will argue that Darwin conceived of species as horizontal similarity complexes demarcated objectively by the law of natural selection. Moreover in the final chapter of my book on the modern species problem (Stamos 2003), I develop an ontology of species as relations by modifying Darwin's ontology to include demarcating causal processes besides natural selection. In all of this I cannot help but find in Lewes a most interesting precursor.

6. This last clause is important, for in spite of his nominalism Lamarck (1809) did allow that what are typically called species "have really only a constancy relative to the duration of the conditions in which are placed the individuals composing it" (36). Thus for

Lamarck species on the traditional view, which involves "an absolute constancy in nature" as well as the belief that "the existence of these species is as old as nature herself" (35), have no real referents in nature. Their reality is conceptual at best. Does that mean that for Lamarck species have a reality in nature in a *different* sense, in particular so long as there are no causal changes acting up them, where "them" shares the traditional definition of "Any collection of like individuals which were produced by others similar to themselves" (35)? This indeed is a possible reading of Lamarck, although it was not the way he was commonly understood.

7. Particularly disappointing on this topic is the recently published anthology *The Cambridge Companion to Darwin* (Hodge and Radick 2003). Not only does it contain next to nothing on Darwin on the ontology of species, but Radick (2003, 165 n. 26) lumps my earlier writings with Beatty, Hodge, and McOuat, no less as sources backing his view that for Darwin naturalists and taxonomists, not nature, "invented" (162) and "divided species from one another" (151), while Gayon (2003, 264 n. 59) uses my earlier writings as the sole reference for his view that Darwin made taxonomic categories, including that of species, "a matter of mere convenience" (260), even though I have never subscribed to the received view.

8. It is interesting to track the effect this line had on Gray. In Gray's review of Darwin's reviewers (Gray 1860c), published for October 1860, although he still ascribes to both Agassiz and Darwin the view that species and varieties exist "as categories of thought" (420–421), to which he adds the word "only" (421), he nevertheless explicitly ascribes to Darwin the view that species in nature are "temporary" (406, 420), and he even goes so far as to claim that "Darwin clearly maintains—what the facts warrant—that the mass of a species remains fixed so long as it exists at all, though it may set off a variety now and then" (424)—a view remarkably similar (though Darwin in fact did not have it) to the modern theory of punctuated equilibria developed and maintained by Eldredge and Gould (1972). Apparently when Gray wrote this paper he had the benefit of Darwin's letter, and thus could no longer, as he did in an earlier review (1860a) discussed near the middle of the present chapter, simply ascribe to Darwin the view that species are "subjective."

CHAPTER 2

1. According to Ghiselin (1989), among good candidates are Mayr's theory of allopatric speciation, according to which most species evolve as the result of a founder effect, which Ghiselin says should be called "Mayr's Law" (58); another is Bergmann's Rule, "which states that in colder climates animals get larger" (61); a further example is Dollo's Law (64), which will be discussed below.

CHAPTER 3

1. Interestingly, this rejected theory of species aging made its way into the first edition of Darwin's *Journal of Researches* (Darwin 1839), but was excised for the second edition (Darwin 1845). In the first edition (1839), at the very end of his first chapter on Patagonia, Darwin wrote "All that at present can be said with certainty, is that, as with the individual,

so with the species, the hour of life has run its course, and is spent" (212). Cf. Gruber (1981, 261, 268–271) for an explanation of why this dated view of Darwin's made it into the first edition.

2. Darwin insisted on gradualism for a number of reasons. In the *Origin* he gives at least six: variation itself is a slow process, niche changes are generally very slow (recall Lyell's uniformitarianism), intercrossing further retards the process (108), saltations (macromutations) are rare (10) and almost always are either maladaptive or nonadaptive (44), saltations do not explain the ubiquitous co-adaptations both within species and between species (3–4), and there is almost no known organ which is not known to have transitional steps (194). It must be added that with his gradualism Darwin was not dogmatic. In reply to the botanist Harvey, who argued that in the Royal Botanic Gardens at Kew a new species of *Begonia* had arisen via saltation, which he called *B. frigida*, Darwin in his correspondence in 1860 (Burkhardt *et al.* 1993) replied to Hooker (February 20) that "Harvey's is a good hit" but that "It would take a good deal more evidence to make me admit that forms have often changed by saltum" (97). To Lyell (February 18–19) he replied that "Harvey does not see that if only a few (as he supposes) of the seedlings inherited his monstrosity natural selection would be necessary to select & preserve them" (93) and that (February 23) "On the whole I still feel excessively doubtful whether such abrupt changes have more than very rarely taken place" (102). And to Harvey himself (September 20–24) he wrote "About sudden jumps; I have no objection to them; they would aid me in some cases: all I can say is, that when I went into the subject, & found no evidence to make me believe in jumps, & a good deal pointing in the other direction—" (373). It is to be noted that in none of his replies does Darwin speak of *B. frigida* as a "species"; instead he speaks of it only as a "form." This is significant, as we shall see in Chapter 5.

3. For examples from Darwin's barnacle monograph that pretty clearly correspond to this passage, cf. Darwin (1854b, 197, 243, 308).

4. Cf. Ospovat (1981, 255–256 n. 3) as well as Darwin in his Historical Sketch (Burkhardt *et al.* 1993), in which he ascribes to Lamarck a belief in "a law of progressive development" (573) and to Chambers a belief in "vital forces" such that "organization progresses by sudden leaps" (574).

CHAPTER 4

1. As for the "competent judges" Darwin mentions in the *Origin*, I have been unable to ascertain their identity. Darwin's section in *Variation* on cattle (Darwin 1868 I, 82–97) is of little help here. The only pre-*Origin* source he mentions is a Memoir by a Professor Nilsson (84 n. 37).

2. It is interesting to note that, contrary to Huxley who told Darwin in a letter (before October 3, 1857) that he thought classification should be irrelevant of pedigree (Burkhardt and Smith 1990, 461), Darwin thought that genealogical classification will someday be achieved more or less. In a letter to Huxley (September 26, 1857) he speculated that "The time will come I believe, though I shall not live to see it, when we shall have fairly true genealogical trees of each great kingdom of nature" (456). Moreover in a letter to Lyell (September 22–23, 1860), Darwin provided two very similar hypothesized genealogical trees for mammals (Burkhardt *et al.* 1993, 379–380).

3. Interestingly, reproductive isolation is in fact not the only consequence of poly-ploidy. In a detailed review of a considerable amount of literature on polyploidy, Levin (1983) concluded that "The biological, physiological, and developmental changes which incidentally accompany chromosome doubling may immediately adapt polyploids to con-ditions other than those to which their diploid progenitors are adapted. In a sense, chromo-some doubling may propel a population into a different resource or habitat space. . . . Polyploidy not only alters the adaptive gestalt of populations, it promotes a series of genetic and chromosomal changes which compound the differences between the polyploid popu-lation and its progenitors. . . . Polyploidy may evoke large, discontinuous effects, which in turn may lead to abrupt, transgressive and manifest shifts in the adaptive gestalt of popula-tions which selection might accomplish slowly or not at all" (15–16). Levin's use of the words "adapt" and "adaptive," of course, are problematic, since in modern evolutionary biology, following Darwin, these terms are normally restricted to products of natural selec-tion (cf. West-Eberhard 1992). We shall return to this issue in the next chapter.

CHAPTER 5

1. Darwin's reasons for concluding common descent are many, and are sprinkled throughout his chapter on the races of man.

2. In *Natural Selection* Darwin provides the example of North American Asters dis-cussed by Torrey and Gray (Stauffer 1975, 196).

3. The only problem with this theory, as I see it, is that Darwin fully recognized that there may be different adaptations *within* the same species. The problem is not so much with the products of sexual selection, as Darwin regarded these as relatively superficial (cf. Dar-win 1859, 88–89, 1871 I, 258–259), much the same as the products of artificial selection compared with the products of natural selection (as we shall see below). Instead, the prob-lem is with the differences within a species that are adaptations produced by natural selec-tion, namely, sexual dimorphism where the differences are to "wholly different habits of life in the two sexes" (cf. Darwin 1859, 87–88, 1871 I, 254–256), caste polymorphism (cf. Dar-win 1859, 236–242), and ontogenetic dimorphism (cf. Darwin 1859, 440, 1868 II, 51–52). Darwin definitely did not divide these differences into different species. Nor did anyone else. As Darwin put it in the *Origin*, "With species in a state of nature, every naturalist has in fact brought descent into his classification; for he includes in the lowest grade, or that of a species, the two sexes; and how enormously these sometimes differ in the most important characters, is known to every naturalist . . . yet no one dreams of separating them. The naturalist includes as one species the several larval stages of the same individual, however much they may differ from each other and from the adult" (424). What all of this indicates, it would seem, is that Darwin's use of the term "form" (e.g., Darwin 1859, 480, 485) had a populational aspect to it. He only used different adaptations to separate different populations into different species, never different sexes, castes, or ontogenetic stages. This is something that we shall see repeated consistently by Darwin in example after example.

4. One can only imagine the shock and sting on Darwin's part, then, after sending Herschel a complimentary copy of the *Origin*, upon hearing that Herschel dismissed it as "the law of higgledy-piggledy," which Darwin heard "by round about channel" and which

he took to be "very contemptuous," adding that "If true this is great blow & discouragement" (Burkhardt and Smith 1991, 423).

5. It is interesting to note that Whewell's response to the *Origin* was much kinder than Herschel's. In a letter to Darwin (January 3, 1860) he wrote "I have to thank you for a copy of your book on the 'Origin of Species.' You will easily believe that it has interested me very much, and probably you will not be surprised to be told that I cannot, yet at least, become a convert to your doctrines. But there is so much of thought and fact in what you have written that it is not to be contradicted without careful selection of the ground and manner of the dissent, which I have not now time for" (Burkhardt *et al.* 1993, 6). In a letter to Lyell written the next day (January 4) Darwin wrote "Possibly you might like to see enclosed note from Whewell, merely as showing that he is not horrified with us" (15).

6. Or rather Darwin *restricted* the term to process laws. In *Variation* Darwin (1868 I) defines laws as "only the ascertained sequence of events" (7). His influence here might be from his reading (either primary or secondary) of Hume. Although there is no evidence that Darwin ever read Hume's *Treatise* or *Enquiry*, we do know that he had read Hume's *Dialogues* (Burkhardt and Smith 1988, 458), which contains a brief (Philo in part 2) of Hume's famous theory of causation as mere regularity, as well as Burton's *Life and Correspondence of David Hume* (472, 479). At any rate, Darwin's expansion (or restriction) of laws to include process laws such as natural selection might help to explain Herschel's rejection of natural selection as "the law of higgledy-piggledy" in note 4 above.

7. Darwin, interestingly, often inveighed against splitters. Since varieties on his view are incipient species and the history of life is treelike (Darwin, again, was a pluralist with regard to speciation), he required what today is called a "polytypic" conception of species (species with varieties or subspecies). As he wrote to Jenyns (April 28, 1858), "One chief reason why I have not accumulated more facts of variation in state of nature is, that naturalists so invariably turn around & say oh they are not varieties, but species" (Burkhardt and Smith 1991, 86). Little wonder, then, that Darwin corresponded so much with lumpers, notably Hooker, Watson, and Gray. For an interesting discussion on the diverse reasons and motivations for Hooker as a lumper, cf. Stevens (1997).

8. Application of the *vera causa* ideal to species classification was not without precedent. For a detailed discussion on, for example, A.P. de Candolle's view on botany, natural classification, and the laws of crystallography, cf. Stevens (1984).

9. Interestingly, when Maw in a letter to Darwin (March 15, 1861) brought out the analogy between mineral classification and organism classification and the problem of superficial similarity between different forms "which look most singularly like genealogical affinity" (Burkhardt *et al.* 1994, 56), Darwin replied (March 17) "I have been particularly struck with your observations on the classification of mineral bodies. The idea has crossed my mind in a very vague & feeble manner: it is so difficult to be honest that I fear that I must have unconsciously banished the idea as disagreeable. I see now the full force of the difficulty & I will make a note not to forget this subject. I must own that classification may be closely like that due to descent & yet have no relation to it" (56–57). Beginning with the fourth edition of the *Origin* Darwin included references to the analogy between mineral and biological classification. In the sixth edition (1876c), for example, on the problem of convergence, he wrote "The shape of a crystal is determined solely by the molecular forces, and it is not surprising that dissimilar substances should sometimes assume the same form;

but with organic beings we should bear in mind that the form of each depends on an infinitude of complex relations, . . . It is incredible that the descendants of two organisms, which had originally differed in a marked manner, should ever afterwards converge so closely as to lead to a near approach to identity throughout their whole organization. If this had occurred, we should meet with the same form, independently of genetic connection, recurring in widely separated geological formations; and the balance of evidence is opposed to any such admission" (100–101). Again he wrote "We know, for instance, that minerals and the elemental substances can be thus arranged ['either artificially by single characters or more naturally by a number of characters']. In this case there is of course no relation to genealogical succession, and no cause can at present be assigned for their falling into groups. But with organic beings the case is different, and the view given above accords with their natural arrangement in group under group; and no other explanation has ever been attempted" (364). Cf. also *Descent* (1871 I, 231) for the same point about convergence and the disanalogy between mineral and organism classification.

Chapter 6

1. To the modern mind, with its emphasis on the genetic basis of sterility, it is perhaps difficult to believe that Darwin would have believed that mere domestication could eventually eliminate such a physical character as sterility between species. Given Darwin's claim in Notebook C (161) that "My definition of species has nothing to do with hybridity, is simply, an instinctive impulse to keep separate, which no doubt be overcome, but until it is the animals are distinct species" (Barrett et al. 1987, 289), combined with his view in the *Origin* that "Natural instincts are lost under domestication" (215), it might be natural to suppose that according to Darwin the domestic breeds of dogs are interfertile because their domestication eliminated the instinct to keep separate between the interfertile wild species from which they were derived. Although a plausible theory on the surface, it nevertheless does not express Darwin's view. Darwin makes this clear enough in the *Origin*, at the end of his discussion on domestic pigs and cattle, in which he wrote "we must look at sterility, not as an indelible characteristic, but as one capable of being removed by domestication" (254).

2. This interpretation finds further corroboration in a letter from Darwin to John Gould, the ornithologist who classified the bird specimens sent to him from Darwin's *Beagle* voyage. In 1861 Gould sent to Darwin a copy of his book *An Introduction to the Trochilidæ, or Family of Humming-Birds* published earlier that year. In that book Gould referred to three species of hummingbirds found on the island of Juan Fernandez and two species of hummingbirds found on the island of Jamaica, stating in each case that no crossing or interbreeding had ever been observed between the species. In a letter to Gould (October 6, 1861), Darwin wrote "I see you allude to the crossing of Birds in a state of nature; I, for one, repudiate this notion" (Burkhardt *et al.* 1994, 295). Gould evidently employed in his species concept reproductive isolation, and Darwin here, quite consistently with everything we have seen and shall further see in this chapter, evidently rejects that concept.

3. Even as early as his *Sketch of 1842* (Darwin 1909), Darwin argued that "sterility, though a usual, is not an invariable concomitant" (50), to which he adduced evidence from Kölreuter, who we know from Darwin's reading notebooks he had read in January of 1840 (Burkhardt and Smith 1988, 461).

4. Interestingly, during 1862 Darwin toyed with, and possibly even held, the idea that interspecific sterility can be naturally selected. His strongest statement to this effect is found in a letter to Hooker (December 12), in which he wrote "By the way my notions on hybridity are becoming considerably altered by my dimorphic work: I am now inclined to believe that sterility is at first a selected quality to keep incipient species distinct. If you have looked at Lythrum, you will see how pollen can be modified merely to favour crossing; with equal readiness it could be modified to prevent crossing" (Burkhardt *et al.* 1997, 598). Darwin quickly gave up this idea, however, and never toyed with it again. Kottler (1985, 402–405) claims that Darwin reverted to his former view the more he studied *Lythrum salicaria*, publishing his results in 1865. Burkhardt *et al.* (1997, 700–711), however, argue that Darwin's *Lythrum* work was part of a wider rejection of selection for sterility, beginning with a plan for breeding experiments on pigeons for the purpose of artificially selecting sterility. In late December 1862 Darwin suggested to his friend, the pigeon fancier Tegetmeier, that he perform the necessary experiments, which Tegetmeier did with negative results (cf. *Variation* 1868 I, 200). Elsewhere in *Variation* (1868 II, 169–172) Darwin gives three reasons for why interspecific sterility could not be selected for, the first being that interspecific sterility could not possibly be selected for in geographically isolated varieties, the second being that selection could not possibly explain the many cases of asymmetry in the sterility/fertility of reciprocal crosses, the third being that it could be of no advantage to individuals of one variety to have reduced fertility with individuals of another variety and thus leave fewer offspring. Burkhardt *et al.* (1997) argue that Darwin wrote the first draft of this section of *Variation* between April and June of 1863 and that "there is no evidence to suggest that the first draft differed from the published version in this regard" (704).

5. It is interesting to compare what Darwin says in the *Origin* on primroses and cowslips with what he says in his 1869 paper. As we have seen in chapter 5, Darwin argued that primroses and cowslips should be classified as two species instead of one and that it is very doubtful that the intermediate links are hybrids. In his later paper on the subject, Darwin (1869) gives essentially the same reasons for classifying primroses and cowslips as different species, but he argues instead that the common oxlip is indeed a hybrid between the two. Nevertheless Darwin nowhere in his paper calls the common oxlip a "species" (which if he did it would create a serious problem for my thesis in chapter 5). Ghiselin, however, is highly misleading in the passage that I quoted at length earlier in the present chapter. He implies that the "third species" is the common oxlip. But it is not. The three species Darwin there refers to are not the primrose, cowslip, and common oxlip, but rather the primrose (*Primula vulgaris*), cowslip (*P. veris*), and Bardfield oxlip (*P. elatior*), the latter which Darwin argues (449–451), much like the former two, is not a hybrid but should be considered a good and true species in its own right.

6. Kottler does indeed seem to think that Darwin's mature species concept is closer to Mayr's than anyone else's (Kottler 1978, 280 n. 9, 292, 297).

7. Of further relevance and interest is what Darwin has to say only a few pages earlier on the origin of "permanent varieties": "The formation of a permanent variety, implies not only that the modifications are inherited, but that they are not disadvantageous, generally that they are in some degree advantageous to the variety, otherwise it could not compete with its parent when inhabiting the same area. The formation of a permanent variety must be effected by natural selection" (Stauffer 1975, 240). To this we should recall to mind Darwin's characterization of species in the *Origin*, as "only well-marked and permanent

varieties" (133), or in other words "only strongly marked and fixed varieties" (155), as well as his claim that varieties are "incipient species" (111).

8. Similar though much abbreviated statements can be found in the *Origin* (101–106).

9. Darwin's passing references in his review paper to examples of preferential mating are apparently taken from his much more careful discussion in *Natural Selection* (Stauffer 1975), in which he gives thirteen examples. Especially significant in his discussion is the example of two forms of Caribou deer, distinguished primarily by different kinds of ranges and migration patterns. Darwin quotes a source as saying that they constitute "two well-marked & permanent varieties" (258).

10. The same point recurs in Darwin's reply to Watson on the topic of convergence at the species level and in his subsequent writings on the topic, as examined in chapter 4. Of further interest is that one can read in Darwin a distinction between adaptations in terms of function and adaptations in terms of structure. In the *Origin* Darwin gives the example of the upland goose, which although it has webbed feet it rarely or never goes near the water, so that "habits have changed without a corresponding change of structure. The webbed feet of the upland goose may be said to have become rudimentary in function, though not in structure" (185). Such structures are nevertheless still adaptations in terms of structure, and are not to be confused with rudimentary organs or structures in the strict sense, which are no longer adaptations in either function or structure because they are atrophied (cf. 416, 450–456, 480, 483).

CHAPTER 7

1. The same might be said of other responses to Darwin's *Origin*. Masters (1860), for example, while maintaining "the existence of 'species,' endowed with a very variable, but a limited power of variation" (224), also maintained that varieties are "confessedly artificial distinctions" (226). Less explicitly, Hopkins (1860) claimed the same when he accused Darwin of confounding "artificial with natural species" (78)—the latter being "formed by nature," the former being manmade and "arbitrary" (747)—in that for Hopkins natural species have a clear distinction from varieties but artificial species do not. Sedgwick (1860) also implied that varieties in nature are not real. According to Sedgwick, "We all admit the varieties, and the very wide limits of variation, among domestic animals," but "the varieties, built upon by Mr. Darwin, are varieties of domestication and human *design*. Such varieties could have no existence in the old world" (285). What I shall argue in subsequent pages, however, is that what we see here in the examples of Agassiz, Masters, Hopkins, and Sedgwick is probably not really a rearguard action in response to the *Origin*, but rather the making explicit, caused by the *Origin*, of a view that had been generally accepted by naturalists prior to the *Origin* but that had remained pretty much implicit until that time.

2. Winsor (1979, 104–111) takes issue with Mayr's claim that Agassiz's variety nominalism "forced" him to split species when there was a lack of intergrading. Instead she claims that prior to the *Origin*, in his *Essay*, Agassiz allowed for the existence of subcategories, including varieties. In all of this, however, her argument, including the example of three- and four-toed box turtles (99–102), as well as the single passage in the *Essay* for subcategories that

she provides as direct evidence (101), does not seem at all to me convincing. (The subcategories briefly discussed by Agassiz are subclasses, suborders, subfamilies, subgenera, and varieties.) The passage she quotes as direct evidence is the following: "The individuals of a species, occupying distinct fields of its natural geographical area, may differ somewhat from one another, and constitute varieties, etc." But it is highly unlikely that Agassiz would have thought that subcategories, including varieties, also belong to God's "mode of thinking." Otherwise what was there to stop the next step of God thinking in terms of evolution? Revealing on this matter is one of Agassiz's early footnotes in his *Essay*, in which he states "It must not be overlooked here that a system may be natural, that is, may agree in every respect with the facts in nature, and yet not be considered by its author as the manifestation of the thoughts of a Creator, but merely as the expression of a fact existing in nature—no matter how—which the human mind may trace and reproduce in a systematic form of its own invention" (8–9 n. 7). This footnote needs to be related to the passage in the *Essay*, which Winsor quotes from, in which Agassiz discusses subcategories. Agassiz specifically states that although they "must be acknowledged in a natural zoological system," they are a matter of "propriety" used "to express all the various degrees of affinity of the different members of any higher natural group" (180), such that, on Agassiz's view, until an objective principle is discovered for the limits of these subdivisions—and he doubts that this will ever happen because they are based on "degrees"—they "must be left to arbitrary estimations," and he hopes that "such arbitrary estimations are forever removed from our science, as far as the categories themselves are concerned" (181). Given all of this, it seems highly doubtful that Agassiz changed his expression on the objective existence of varieties following the publication of the *Origin*. He seems never to have believed in their objective existence. Instead his only change was a change in emphasis.

3. According to Hodge (1987), Linnaeus is largely to be credited for laying the foundation for "the historical and teleological conception of 'permanent varieties' that has been developed by Prichard's time" (232), but what we have seen already from Linnaeus above undermines to my mind that assessment.

4. This letter may well be the source for an interesting passage in *Natural Selection*, in which Darwin wrote "Dr. Hooker objects to my whole manner of treating the present subject because varieties are so ill defined; had he added that species were likewise ill defined, I should have entirely agreed with him; for my belief is that both are liable to this imputation; varieties more than closely allied species, & these more than strongly marked species" (Stauffer 1975, 159–160). We shall see the significance of this further below.

CHAPTER 8

1. For a discussion on Aristotle's essentialism, with his distinction between *genos* and *eidos* (which is not to be confused with the modern taxonomic distinction between genus and species), cf. Stamos (2003, 102–111) and Grene and Depew (2004, 10–34). Ghiselin's explicit source for his understanding of Aristotelian essentialism and categories is Hull (1965), who himself did not use primary but only secondary sources.

2. This is indicated by the fact that it has been reprinted in an anthology devoted to the species problem, namely, Ereshefsky (1992a), in the direct following that its line of interpretation has gained (e.g., Bowler 1988, 69; Stevens 1992, 305; Kitcher 1993, 32 n. 45;

de Queiroz 1999, 83 n. 24; Grene and Depew 2004, 213 n. 20; LaPorte 2004, 192 n. 13), and in the slight twists that have been added to it by others (e.g., Hodge 1987, 248–249; McOuat 1996, 475, 514–515).

3. This theory isn't entirely original. Ellegård (1958), for example, claimed that "Varietal distinctions were accidental and fluctuating, while specific ones were essential and permanent. Thus the immutability doctrine was, as it were, guaranteed by the current definition of species, and this was one of the first obstacles that Darwin had to remove in order to prepare the way for his transmutation theory" (198). A few years later Simpson (1961) claimed that "What he [Darwin] did was to take as given the classifications then current and to show, first, that they were consistent with the theory that their taxa originated by evolution, and second, that evolutionary phylogeny could explain the order that had *already* been found among organisms" (53).

4. To all of this one must keep in mind that Darwin began his transmutation notebooks in July 1837 (Barrett *et al.* 1987, 167; cf. Darwin 1876a, 83).

5. "With species in a state of nature, every naturalist has in fact brought descent into his classification; for he includes in his lowest grade, or that of a species, the two sexes; and how enormously these sometimes differ . . . yet no one dreams of separating them. The naturalist includes as one species the several larval stages of the same individual, however much they may differ from each other and from the adult; . . . He includes monsters; he includes varieties, not solely because they closely resemble the parent-form, but because they are descended from it" (Darwin 1859, 424).

6. "Several eminent naturalists have of late published their belief that a multitude of reputed species in each genus are not real species; but that other species are real, that is, have been independently created" (Darwin 1859, 482).

7. "Hence I believe a well-marked variety may be justly called an incipient species; but whether this belief be justifiable must be judged of by the general weight of the several facts and views given throughout this work" (Darwin 1859, 52).

8. Hooker let Darwin see this letter to him. Interestingly, in response to Lyell's fear quoted above, Darwin wrote to Hooker (July 30, 1856) "I differ from him greatly in thinking that those who believe that species are *not* fixed will multiply specific names: I know in my own case my most frequent source of doubt was whether others would not think this or that was a God-created Barnacle & surely deserved a name. Otherwise I sh[d]. only have thought whether the amount of difference & permanence was sufficient to justify a name" (Burkhardt and Smith 1990, 194).

9. Burkhardt *et al.* (1993) point out that, according to Francis Darwin, his father was in the habit of periodically burning accumulated letters, such that "This process, carried on for years, destroyed nearly all letters received before 1862" (110). This would explain why in Darwin's extant correspondence prior to 1862 more of the letters are from Darwin than to Darwin.

10. The closest reference that one can find is Hooker's letter to Darwin (March 14, 1858), in which Hooker wrote "The long & short of the matter is, that Botanists do not attach that *definite* importance to varieties that you suppose" (Burkhardt and Smith 1991, 49).

11. According to Burkhardt and Smith (1989), as of mid-1855 "both Joseph Dalton Hooker and Louis Agassiz regarded species to be fixed, but Hooker thought all

242 Notes to Chapter 8

species originated from a single parent or pair whereas Agassiz believed in multiple creation" (364 n. 6).

12. Indeed it was typical of Darwin to mean by "given up species" no more than the rejection of the belief that species are immutable, that is, that one species can evolve into another. For example, in his reading notebooks Darwin wrote of Watson (1845a) that he "gives up permanent species" (Burkhart and Smith 1988, 449). We have already seen in chapter 1 that even though Watson was an early evolutionist he never believed that species are not real. Moreover Darwin never believed him to have taken such a nominalist position.

13. This is in spite of having written in the *Origin*, after listing a number of eminent paleontologists and geologists who do not doubt that species are immutable, that "I have reason to believe that one great authority, Sir Charles Lyell, from further reflexion entertains grave doubts on this subject" (310).

14. An interesting and contemporaneous use of the word "entity" is to be found in Bentham (1858): "He [Linnaeus] accordingly, treating his genera as *entities* (to use a word of Jeremy Bentham's) as natural as species, distributed them for practical purposes into his well-known artificial Classes and Orders" (30). The connotation of realism in this passage is pretty clear, so that a nonentity is not real. Darwin's use of the term "entity" in the passage above, then, might seem to imply species nominalism, but it doesn't necessarily. Darwin's use of the term "entity" only applies to the sterility criterion of species distinctness, for Darwin immediately continues in the above passage to say "Is it not that the dog case injures the argument from fertility; so that one main argument that the races of man are varieties & not species, i.e. because they are fertile inter se is much weakened?" (397).

15. This is a prime example of the kind of reasoning that has become popular in professional history of science, which I take on in chapter 10. The present case, however, deserves a few comments here. First, no reference is given for the "no jot of 'demonstrative evidence in its favour'" line, which is given to imply that Huxley converted for non-evidential reasons, and which is then contradicted by the reference to Darwin on pigeon breeds, a matter of evidence that Darwin surely would have raised in his discussions with Huxley and others at his home. Second, Desmond (675 n. 24) refers to Lyell's journal on the species question (Wilson 1970) as historical support. If we look at what Lyell actually wrote, however, all he wrote in his journal (dated April 29, 1856) is that "After conversation with Mill, Huxley, Hooker, Carpenter & Busk at Philos. Club, conclude that the belief in species as permanent, fixed & invariable, & as comprehending individuals descending from single pairs or protoplasts is growing fainter—no very clear creed to substitute" (56–57). This is hardly good evidence for an "about-face" on Huxley's or anyone else's part. Third and most importantly, Huxley had been a longtime friend of Herbert Spencer, the evolutionary progressivist *par excellence*. Had Huxley's main motive for converting to evolutionism been political, the main catalyst should have been Spencer, not Darwin, and it should have occurred much sooner. A much more plausible explanation, then, would be that it was scientific evidence and argument, raised mainly by Darwin and lacking in Spencer, which got Huxley turned around, evidence and argument that took a couple of years for him to digest and that he subsequently used in his campaign to professionalize science. To this it is interesting to add Huxley's own reflections on the matter recorded many years later. Noting that he first met Spencer in 1852 and that shortly afterward they had become friends, he adds that "Many and prolonged were the battles we fought on this topic [evolution]. But even my friend's rare dialectic skill and copiousness of apt illustration could not drive me from my

agnostic position. I took my stand upon two grounds: firstly, that up to that time, the evidence in favour of transmutation was wholly insufficient; and, secondly, that no suggestion respecting the causes of the transmutation assumed, which had been made, was in any way adequate to explain the phenomena. Looking back at the state of knowledge at that time, I really do not see that any other conclusion was justifiable" (Darwin 1892, 178). In what I call "the new historiography," all of this means nothing.

16. Cf. the correspondence between Darwin and Huxley (late September 1857) on the importance of genealogy versus structure in classification (Burkhardt and Smith 1990, 456, 461–462).

17. It might seem odd that Darwin would read Wallace (1855) as a creationist, and continued to do so for some time. For one thing, Lyell himself never did, or did so only hesitatingly. In his index book to his first journal on the species question, begun on November 28, 1855, Wallace's article is the first topic of writing, and Lyell wrote that Wallace "goes far towards Lamarck's doctrine" (Wilson 1970, 66). During a visit to Darwin's home from April 13–16, 1856, Lyell apparently recommended Wallace's paper for Darwin to read and warned him that he might be forestalled (Burkhardt and Smith 1991, 108 n. 4). What may have contributed to Darwin's reading of Wallace (1855) as a creationist, in spite of Lyell, is that Wallace repeatedly used the word "creation" (188, 193), he repeatedly referred to the closely-allied ancestral species in his "law" as the "anti-type" (186, 187, 188, 191, 192), and even though he used the tree metaphor (187, 191) he thought that "We have no reason for believing that the number of species on the earth at any former period was much less than at present" (194). Thus even though Wallace used the tree metaphor, argued for a kind of gradualism in change between species (191, 195–196), and argued for the importance and significance of rudimentary organs (195), Darwin read Wallace as a successive creationist, which indeed at the time was nothing new. Many, such as Lyell and Agassiz, had combined the *natura non facit saltum* creationism of Linnaeus with the idea of many successive creations.

CHAPTER 9

1. For his maintenance of "generally," cf. Kuhn (1970, 181) and Hoyningen-Huene (1993, 232–233).

2. Cf., for example, Lakatos (1970, 93) and Chalmers (1999, 115, 123–124) for a reading of Kuhn as meaning religious conversion, and Hoyningen-Huene (1993, 257–258, 239–252) for a disavowal that religious conversion was what Kuhn had meant.

3. Kuhn was quite apparently an instrumentalist, as, for example, when he (Kuhn 1970) compares scientific theories to tools and says "As in manufacture so in science—retooling is an extravagance reserved for the occasion that demands it. The significance of crisis is the indication they provide that an occasion for retooling has arrived" (76).

4. On the topic of age as a factor, Hull *et al.* (1978), restricting themselves to the Darwinian revolution, claim that "age explains less than ten percent of the variation in acceptance," so that "the connection between age and acceptance is not as important as people such as Max Planck have claimed" (58). Sulloway (1996) calls age "a reasonably good predictor of attitudes toward scientific innovation" (36), but adds the qualification that

with firstborns age does not seem to play much of a factor, whereas with laterborns it "exerts a substantially greater influence on attitudes toward evolution" (35). Darwin, for what it's worth, thought that age would be a large factor. In a passage surprisingly quoted with approval by Kuhn (1970, 151), Darwin wrote at the end of the *Origin* that he looked mainly to "young and rising naturalists" to transform creationist biology into evolutionary biology, since he thought they "would be able to view both sides with impartiality" (482). This was a theme that Darwin would repeat throughout his correspondence (cf. Burkhardt and Smith 1991, 279, 404, 431; Burkhardt *et al.* 1993, 115, 507, 514; Burkhardt *et al.* 1994, 45, 102, 135).

5. Originally Kripke and Putnam argued that what unites samples of a natural kind is an underlying structural essence, whether chemical kinds such as gold (the essence being proton number 79) or biological kinds including species such as tiger (the essence being in the chromosome structure). Whether the causal theory of reference is necessarily committed to essentialism for natural kinds (which of course is utterly implausible for biological species, since variation is the norm) is discussed and rejected, for example, by Hacking (1983, 82).

6. As Putnam (1973) put it, "concepts which are not strictly true of anything may yet refer to something; and concepts in different theories may refer to the same thing . . . realists have held that there are successive scientific theories about the *same* things: about heat, about electricity, about electrons, and so forth; and this involves treating such terms as 'electricity' as *trans-theoretical* terms" (197).

7. Again as Putnam (1973) put it, "Scientists . . . are trying to maximize *truth* (or improve their approximation to truth, or increase the amount of approximate-truth they know without decreasing the goodness of the approximation, and so forth)" (212).

8. Beatty in his writings about Darwin on species makes no *explicit* reference to the causal theory of meaning; indeed he coyly hides his debt to it, claiming that his discussion is in terms of nineteenth century thought. Nevertheless one clue to his debt is that he approvingly refers to Kitcher's (1978) discussion on "recent developments in semantics" and claims that Darwin did indeed recognize and take advantage of the distinction between meaning as reference and meaning as definition (cf. Beatty 1982, 221–222 n. 6, 1985, 280–281 n. 3).

9. Bowler (2000) claims that Darwin's conversion was "like a gestalt-shift" (95), that Darwin "jumped suddenly and wholeheartedly" (97), but this seems to me not only to ignore the evidence to the contrary but to be the result of a confusion. Granted, in his autobiography Darwin (1876a) states that the theory of natural selection "at once struck" him (120) in October 1838 after reading Malthus, and also that he can remember "the very spot in the road, whilst in my carriage" (120) at which the principle of divergence "occurred" to him many years later, but it does not follow that his *conversion* to these theories was saltational rather than gradual. For one, he also states that it was not until "about 1839" that "the theory [of evolution by natural selection] was clearly conceived" (124). For another, his principle of divergence was a theory that he subsequently went to great lengths to test (as we shall see in the next chapter).

Chapter 10

1. This is, ironically and paradoxically, an epistemological premise, a statement about reality, specifically people including scientists, which is claimed to be true, and yet it is

thoroughly immune to testing—since by its own principles no amount of evidence could logically refute it!

2. The Copernican revolution, it should be noted, did not follow immediately after the publication of Copernicus' *De Revolutionibus*, but instead took roughly 150 years to be completed (Kuhn 1957).

3. It is interesting that Darwin, a few years after the publication of the *Origin*, claimed that the most important change was to the belief in species evolution, not to natural selection, although he did think that the latter would eventually be accepted. In a letter to Hooker (January 13, 1863), for example, he wrote "as I look at it, the great gain is for any good man to give up immutability of species: the road is then open for progress; it is comparatively immaterial whether he believes in N. Selection; but how any man can persuade himself that species change unless he sees how they become adapted to their conditions is to me incomprehensible" (Burkhardt *et al.* 1999, 36). Similarly in a letter to A. de Candolle (January 14, 1863) he wrote "But the great point, as it seems to me, is to give up the immutability of specific forms; as long as they are thought immutable, there can be no real progress in 'epiontology' [the attempt to unify biogeography and phylogeny]" (40). One can also find the same basic claim in one of his articles (Darwin 1863b), in which he wrote "Whether the naturalist believes in the views given by Lamarck, or Geoffroy St.-Hilaire, by the author of the 'Vestiges,' by Mr. Wallace and myself, or in any other such view, signifies extremely little in comparison with the admission that species have descended from other species and have not been created immutable; for he who admits this as a great truth has a wide field opened to him for further inquiry. I believe, however, from what I see of the progress of opinion on the Continent, and in this country, that the theory of Natural Selection will ultimately be adopted, with, no doubt, many subordinate modifications and improvements" (Barrett 1977 II, 81). It should be added that Darwin never gave up his belief in the importance of adaptation and natural selection, as indicated by what he wrote in his autobiography (1876a). On the topic of adaptation, he wrote "I had always been much struck by such adaptations, and until these could be explained it seemed to me almost useless to endeavour to prove by indirect evidence that species have been modified" (199). This needs to be compared with what he wrote on the topic of natural selection: "The old argument of design in nature, as given by Paley, which formerly seemed to me so conclusive, fails, now that the law of Natural Selection has been discovered" (87).

4. To be specific, Agassiz confined recapitulation to each of Cuvier's *embranchements* (cf. Ospovat 1981, 135).

5. Darwin's correspondence is often useful for finding clarifications and emphases perhaps missing in his published work. For example, in a letter to Lyell (March 12, 1860) he wrote that his view of progress depends "on the conditions" (Burkhardt *et al.* 1993, 128), while in a letter to Harvey (September 20–24, 1860) he wrote "I consider Natural Selection of such high importance, because it accumulates successive variations in *any* profitable direction; & thus adapts each new being to its complex conditions of life" (371, italics mine). Equally important, in a letter to Falconer (October 1, 1862) he wrote "This [adaptation] seemed to me and does still seem the problem to solve, and I think natural selection solves it, as artificial selection solves the adaptation of domestic races for man's use. But I suspect that you mean something further,—that there is some unknown law of evolution by which species *necessarily* change; and if this be so, I cannot agree" (Burkhardt *et al.* 1997, 441). Similarly he wrote in a letter to Scott (March 6, 1863) "I cannot help doubting from

your expression of an 'innate . . . . . selective principle' whether you fully comprehend what is meant by Natural Selection" (Burkhardt *et al.* 1999, 213). Cf. also *Natural Selection* (Stauffer 1975, 238, 271, 273), *Variation* (1868 II, 425–428), and Darwin's autobiography (1876a, 87).

6. Sulloway (1996) calls the historiography of Desmond and Moore "Marxist" (240). Moreover, as a result of his enormous statistical analysis that he conducted over many years, Sulloway argues, against Desmond (1989), that "the correlation between social class and support for evolution is almost zero" (237), so that "The idea that social class explains something about the Darwinian revolution is an illusion" (240). He argues further that Desmond's analysis of social class and acceptance of evolution "is the product of biased sampling that confirmed his working hypothesis" (239). I would argue that bias is precisely what characterizes Desmond and Moore (1992). Indeed a few years later, looking back at the co-authoring, Moore (1996) confides that "our historiographic aim was shared: to embed Darwin the man, his practices and theories, in a shifting social order" (275), a historiography that sees "progress" in "the degree to which the pure waters of Darwin's science are muddied by the rich surrounding soil of political economy, natural theology, urban radicalism and provincial Dissent" (271). Mud is exactly what we find (cf. Moore 1997 who does the same for Wallace).

7. Referring to Darwin's seed experiments on *Lythrum salicaria* (1865), in which, as we have seen in chapter 6, Darwin found sterility in homomorphic though not heteromorphic unions, in what he called "illegitimate" as opposed to "legitimate unions" (120), Desmond and Moore (1992) remark that "Sterile seed counts somehow fitted an unromantic, data-crunching age" (520). Although they obviously prefer to substitute Darwin's word "unions" above with "marriages," they nevertheless miss the point about Darwin's species, which is that they undermined the Victorian value of family, with lots of children gathered round the hearth. Indeed, as Secord (1989) points out, it was partly because Chambers in his *Vestiges* (1844) played into the hands of this Victorian domestic virtue—with his generative model of parental and offspring species, such that "the birth of a new species was no more or less to be feared than the birth of a child" (184)—that his book became a Victorian best-seller.

8. It is to be noted that this was indeed part of Darwin's family heritage. Desmond and Moore (1992) refer to the "bastardizing experiments" (520) of his grandfather, Erasmus Darwin (an early evolutionist, it should be added, famous author of *Zoönomia*), while Browne (1995) points out that "Erasmus Darwin had a passionate and earthy nature, at one with the liberal sexual views of eighteenth-century Paris" (41). Charles Darwin's species, then, rather than being good *Victorian* species, were good *Parisian* species! All the more odd that the *Origin* never took off in France.

9. Or rather he was one of the first, since the nod for the basic idea should probably go to Patrick Matthew, who published his theory in the appendix to his *On Naval Timber and Arboriculture* (cf. Burkhardt *et al.* 1993, 584–589), published in 1831. And yet credit must still go to Darwin for being the first in history to give a *scientific* explanation of adaptations by natural selection, which included not only a wealth of evidence but also recognition and explanation of co-adaptations, such as moths and orchids. Striking is his prediction concerning the Madagascar orchid *Angraecum sesquipedale*, with its nectar receptacle "eleven and a half inches long, with only the lower inch and a half filled with nectar" (Darwin 1862a, 162). Darwin predicted that, in order to harvest the nectar, "there must be moths with proboscides capable of extension to a length of between ten and eleven inches!" (163). The discovery of the moth was made 40 years later in Madagascar, long after Darwin was

dead (cf. Browne 2002, 178; interestingly the prediction and discovery are omitted from Desmond and Moore 1992).

10. The immediate ancestor of Winsor's article is Camardi (2001), between which there is much overlap, notably the rejection of the interpretations of essentialism by Hull and Mayr, and the acceptance of the clusterism of Whewell and Boyd. But Camardi presents a different angle than Winsor. Based mainly on his study of the work of von Baer and Owen, he states that "if we read in sequence von Baer, Owen and Darwin, we understand that the Darwinian concept of evolution was not born full-fledged merely from Darwin's own speculation. Rather, it is a result of the long historical accumulation on such concepts as development, generation, reproduction, a work that involved both Owen and Darwin, among others" (500). Here again we see an example of the new historiography, which sees Darwin's theories as the result of a "long historical accumulation," and in which we find phrases such as "must have" (495), "should be" (496), and "sounds like" (509) used to convince us that von Baer and Owen were already testing the evolutionary waters prior to Darwin's coming out, an interpretation that is stretched at best. On the other hand, there is something in Grene and Depew's (2004) claim that "thanks to Darwin, we think of homology as a historical concept. It is difficult for us to think of it in pre-Darwinian, non-historical terms" (215). Of course it is easy for *us* (and professional historians of biology are apparently no exception) to see in von Baer's law of branching development in embryos, or in Owen's work on archetypes and homologies, theories leading to the Darwinian view. We can just as easily see the common belief in the Great Chain of Being, whether the older linear version, or the later branching version established by Cuvier's theory of *embranchements*, or the doctrine of successive creations, paving the way for Darwin. But we seriously need to remember that, unlike for us today, it was quite easy, natural, and common for biologists at the time of Darwin and before to accept the above theories and still view nature in static terms. This was not only because their biology was mixed with religion, with its creationism, but also because they had a belief in species essentialism and in related laws such as the laws of reversion and limited variation. These were enormous obstacles that Darwin needed to overcome, and there were many more. Not surprisingly, von Baer rejected evolution, and pathetically little evidence of evolutionary thinking on the part of Owen exists before 1859. Indeed most of Camardi's evidence for Owen as a moderate evolutionist and "forerunner of Darwin" is post-1859, which is entirely irrelevant. But such is the new historiography. It goes out of its way, seemingly obsessed, to find and overemphasize anything that takes away from Darwin's originality and achievement, while it ignores or underemphasizes anything that supports it. So far, Darwin's species concept has suffered the same fate.

11. It is pernicious not only for its treatment of scientific epistemology but also (relatedly) for its treatment of individual scientists. For a good example of how the new historiography can be pernicious in both ways, cf. Moore (1997), where the rejection of, even the intense dislike for, the concepts of scientific genius and scientific discovery is quite apparent (cf. esp. 295, 307), and where the historiography that views the most basic ideas of scientists as projections from society and culture is forwarded as progress (290–292).

References

All references are to reprints where indicated

Agassiz, Louis (1857). *Essay on Classification*. Boston: Little, Brown & Co. (1859). 2nd ed. London: Longman, Brown, Green, Longmans and Roberts. Reprinted in Edward Lurie, ed. (1962). Cambridge: Harvard University Press.

——— (1860a). "Minutes of Meeting of March 13." *Proceedings of the American Academy of Arts and Sciences* 4, 410.

——— (1860b). "On the Origin of Species." *American Journal of Science and Arts* 30 (2nd ser.), 142–154.

Alter, Stephen G. (1999). *Darwinism and the Linguistic Image: Language, Race, and Natural Theology in the Nineteenth Century*. Baltimore: Johns Hopkins.

Andersson, Malte (1994). *Sexual Selection*. Princeton: Princeton University Press.

Anon. (1959). "Darwin's Origin of Species." *The Saturday Review* Dec. 24, 775–776.

Anon. (1860a). "Professor Owen on the Origin of Species." *The Saturday Review* May 5, 573–574.

Anon. (1860b). "Species." *All the Year Round* June 2, 174–187; July 7, 293–299.

Armstrong, D.M. (1983). *What is a Law of Nature?* Cambridge: Cambridge University Press.

Arthur, J.C. (1908). "The Physiological Aspect of the Species Question." *American Naturalist* 42, 243–248.

Ashton, Paul A., and Abbott, Richard J. (1992). "Multiple Origins and Genetic Diversity in the Newly Arisen Allopolyploid Species, *Senecio Cambrensis* Rosser (Compositae)." *Heredity* 68, 25–32.

Babinski, Edward T. (1995). *Leaving the Fold: Testimonies of Former Fundamentalists*. Amherst, NY: Prometheus Books.

Barrett, Paul, ed. (1977). *The Collected Papers of Charles Darwin*. Two volumes. Chicago: University of Chicago Press.

Barrett, Paul H., *et al.*, eds. (1987). *Charles Darwin's Notebooks, 1836–1844*. Ithaca: Cornell University Press.

Bates, Henry Walter (1862). "Contributions to an Insect Fauna of the Amazon Valley. Lepidoptera: Heliconidæ." *Transactions of the Linnean Society* 23, 495–566.

Baum, David A., and Shaw, Kerry L. (1995). "Genealogical Perspectives on the Species Problem." In Peter C. Hoch and A.G. Stephenson, eds. (1995, 289–303). *Experimental and Molecular Approaches to Plant Biosystematics*. St. Louis: Missouri Botanical Gardens.

Beatty, John (1982). "What's In a Word?: Coming to Terms in the Darwinian Revolution." *Journal of the History of Biology* 15, 215–239.

—— (1985). "Speaking of Species: Darwin's Strategy." In Kohn (1985a, 265–281).

Beer, Gillian (1989). "Darwin and the Growth of Language Theory." In John Christie and Sally Shuttlesworth, eds. (1989, 152–170). *Nature Transfigured: Science and Literature, 1700–1900*. Manchester, UK: Manchester University Press.

Bentham, George (1836). *Labiatarum Genera et Species*. London: James Ridgway and Sons.

—— (1858). "Memorandum on the Principles of Generic Nomenclature in Botany." *Journal of the Proceedings of the Linnean Society (Botany)* 2, 30–33.

—— (1861) "On the Species and Genera of Plants, Considered With Reference to Their Practical Application to Systematic Botany." *Natural History Review* 1, 133–151.

Bessey, Charles E. (1908). "The Taxonomic Aspect of the Species Question." *American Naturalist* 42, 218–224.

Blyth, Edward (1835). "An Attempt to Classify the 'Varieties' of Animals." *Magazine of Natural History* 8, 40–53.

Bowen, Francis (1860). "Darwin on the Origin of Species." *North American Review* 90, 474–506.

Bowler, Peter J. (1988). *The Non-Darwinian Revolution: Reinterpreting a Historical Myth*. Baltimore: Johns Hopkins.

—— (2000). "Philosophy, Instinct, Intuition: What Motivates the Scientist in Search of a Theory?" *Biology & Philosophy* 15, 93–101.

Boyd, Richard (1999). "Homeostasis, Species, and Higher Taxa." In Wilson (1999, 141–185).

Browne, Janet (1995). *Charles Darwin: Voyaging*. New York: Knopf.

—— (2002). *Charles Darwin: The Power of Place*. New York: Knopf.

Burkhardt, Frederick, and Smith, Sydney, eds. (1985). *The Correspondence of Charles Darwin, Volume 1, 1821–1836*. Cambridge: Cambridge University Press.

—— (1986). *The Correspondence of Charles Darwin, Volume 2, 1837–1843*. Cambridge: Cambridge University Press.

—— (1987). *The Correspondence of Charles Darwin, Volume 3, 1844–1846*. Cambridge: Cambridge University Press.

—— (1988). *The Correspondence of Charles Darwin, Volume 4, 1847–1850*. Cambridge: Cambridge University Press.

———— (1989). *The Correspondence of Charles Darwin, Volume 5, 1851–1855*. Cambridge: Cambridge University Press.

———— (1990). *The Correspondence of Charles Darwin, Volume 6, 1856–1857*. Cambridge: Cambridge University Press.

———— (1991). *The Correspondence of Charles Darwin, Volume 7, 1858–1859*. Cambridge: Cambridge University Press.

Burkhardt, Frederick, *et al.*, eds. (1993). *The Correspondence of Charles Darwin, Volume 8, 1860*. Cambridge: Cambridge University Press.

———— (1994). *The Correspondence of Charles Darwin, Volume 9, 1861*. Cambridge: Cambridge University Press.

———— (1997). *The Correspondence of Charles Darwin, Volume 10, 1862*. Cambridge: Cambridge University Press.

———— (1999). *The Correspondence of Charles Darwin, Volume 11, 1863*. Cambridge: Cambridge University Press.

———— (2001). *The Correspondence of Charles Darwin, Volume 12, 1864*. Cambridge: Cambridge University Press.

———— (2002). *The Correspondence of Charles Darwin, Volume 13, 1865*. Cambridge: Cambridge University Press.

Burkhardt, R.W., Jr. (1987). "Lamarck and Species." In Roger and Fischer (1987, 161–180).

Camardi, Giovanni (2001). "Richard Owen, Morphology and Evolution." *Journal of the History of Biology* 34, 481–515.

Carpenter, William Benjamin (1860). "Darwin on the Origin of Species." *National Review* 10, 188–214.

Chalmers, A.F. (1999). *What Is This Thing Called Science?* 3rd ed. Indianapolis: Hackett.

Chambers, Robert (1844). *Vestiges of the Natural History of Creation*. London: John Churchill.

———— (1859). "Charles Darwin on the Origin of Species." *Chambers's Journal* 12, 388–391.

Claridge, M.F., Dawah, H.A., and Wilson, M.R., eds. (1997). *Species: The Units of Biodiversity*. London: Chapman & Hall.

Coleman, William (1962). "Lyell and the 'Reality' of Species: 1830–1833." *Isis* 53, 325–338.

Coulter, J.M. (1908). "Discussion of the Species Question." *American Naturalist* 42, 272–281.

Cowan, S.T. (1962). "The Microbial Species—A Macromyth?" *Symposium—Society for General Microbiology* 12, 433–455.

Cowles, H.C. (1908). "An Ecological Aspect of the Conception of Species." *American Naturalist* 42, 265–271.

Cracraft, Joel (1983). "Species Concepts and Speciation Analysis." *Current Ornithology* 1, 159–187. Reprinted in Ereshefsky (1992a, 93–120).

——— (1987). "Species Concepts and the Ontology of Evolution." *Biology & Philosophy* 2, 329–346.

——— (1989). "Species as Entities of Biological Theory." In Ruse (1989, 31–52).

——— (1997). "Species Concepts in Systematics and Conservation Biology—An Ornithological Viewpoint." In Claridge *et al.* (1997, 325–339).

Crawford, Charles, and Krebs, Dennis L., eds. (1998). *Handbook of Evolutionary Psychology: Ideas, Issues, and Applications.* Mahwah, NJ: Lawrence Erlbaum Associates.

Crawfurd, John (1859). "On the Origin of Species." *The Examiner* Dec. 3, 772–773.

Creath, Richard, and Maienschein, Jane, eds. (2000). *Biology and Epistemology.* Cambridge: Cambridge University Press.

Cronquist, Arthur (1978). "Again, What Is a Species?" *Biosystematics in Agriculture: Beltsville Symposia in Agricultural Research* 2, 3–20.

Dana, James D. (1857). "Thoughts on Species." *Annals and Magazine of Natural History* 20 (2nd ser.), 485–497.

Darwin, Charles (1839). *Journal of Researches into the Geology and Natural History of the Various Countries Visited by H.M.S. Beagle, Under the Command of Captain Fitzroy, R.N. from 1832 to 1836.* London: Henry Colburn.

——— (1845). *Journal of Researches into the Natural History and Geology of the Various Countries Visited During the Voyage of H.M.S. Beagle Round the World.* 2nd ed. London: John Murray.

——— (1851a). *A Monograph on the Fossil Lepadidæ.* London: The Paleontological Society.

——— (1851b). *A Monograph on the Sub-Class Cirripedia, the Lepadidæ.* London: The Ray Society.

——— (1854a). *A Monograph on the Fossil Balanidæ.* London. The Paleontological Society.

——— (1854b). *A Monograph on the Sub-Class Cirripedia, the Balanadæ.* London: The Ray Society.

——— (1858). "On the Agency of Bees in the Fertilization of Papilionaceous Flowers, and on the Crossing of Kidney Beans." *Annals and Magazine of Natural History* 2 (3rd ser.), 459–465. Reprinted in Barrett (1977 II, 19–25).

——— (1859). *On the Origin of Species by Means of Natural Selection.* London: John Murray.

——— (1862a). *The Various Contrivances by which Orchids are Fertilised by Insects.* London: John Murray. 2nd ed. (1877).

——— (1862b). "On the Two Forms, or Dimorphic Condition, in the Species of *Primula*, and on Their Remarkable Sexual Relations." *Journal of the Proceedings of the Linnean Society (Botany)* 6, 77–96. Reprinted in Barrett (1977 II, 45–63).

——— (1863a). "The Doctrine of Heterogeny and Modification of Species." *Athenaeum* 9, 554–555. Reprinted in Barrett (1977 II, 78–80).

——— (1863b). "Origin of Species." *Athenaeum* 9, 617. Reprinted in Barrett (1977 II, 81).

——— (1863c). "A Review of H.W. Bates' Paper on 'Mimetic Butterflies.'" *Natural History Review* 3, 219–224. Reprinted in Barrett (1977 II, 87–92).

——— (1865). "On the Sexual Relations of the Three Forms of *Lythrum salicaria.*" *Journal of the Proceedings of the Linnean Society (Botany)* 8, 169–196. Reprinted in Barrett (1977 II, 106–131.

——— (1868). *The Variation of Animals and Plants Under Domestication.* Two volumes. London: John Murray. 2nd ed. (1875).

——— (1869). "On the Specific Difference Between *Primula veris*, Brit. Fl. (var. *acaulis*, Linn.), and *P. elatior*, Jacq.; and on the Hybrid Nature of the Common Oxlip. With Supplementary Remarks on Naturally-Produced Hybrids in the Genus *Verbascum.*" *Journal of the Linnean Society (Botany)* 10, 437–454.

——— (1871). *The Descent of Man, and Selection in Relation to Sex.* Two volumes. London: John Murray.

——— (1872). *The Expression of the Emotions in Man and Animals.* London: John Murray.

——— (1876a). *Autobiography.* In Nora Barlow, ed. (1958). *The Autobiography of Charles Darwin, 1809–1882.* London: Collins.

——— (1876b). *The Effects of Cross and Self Fertilisation in the Vegetable Kingdom.* London: John Murray. 2nd ed. (1878).

——— (1876c). *The Origin of Species.* 6th ed. London: John Murray.

——— (1880). "Fertility of Hybrids from the Common and Chinese Goose." *Nature* 21, 207. Reprinted in Barrett (1977 II, 219–220).

Darwin, Francis, ed. (1892). *Charles Darwin: His Life Told in an Autobiographical Chapter and in a Selected Series of his Published Letters.* New York: Appleton.

———, ed. (1909). *The Foundations of the Origin of Species: Two Essays Written in 1842 and 1844.* Cambridge: Cambridge University Press.

Davies, Anna Morpurgo (1987). "'Organic' and 'Organism' in Franz Bopp." In Hoenigswald and Wiener (1987, 81–107).

Dawkins, Richard (1983). "Universal Darwinism." In D.S. Bendall, ed. (1983, 403–425). *Evolution from Molecules to Men.* Cambridge: Cambridge University Press.

——— (1986). *The Blind Watchmaker.* Harlow, UK: Longman Scientific & Technical.

Dawson, John William (1860). "Review of 'Darwin on the Origin of Species by means of Natural Selection.'" *Canadian Naturalist* 5, 100–120.

de Queiroz, Kevin (1999). "The General Lineage Concept of Species and the Defining Properties of the Species Category." In Wilson (1999, 49–89).

Dennett, Daniel (1995). *Darwin's Dangerous Idea: Evolution and the Meanings of Life.* New York: Simon & Schuster.

Desmond, Adrian (1989). *The Politics of Evolution: Morphology, Medicine, and Reform in Radical London.* Chicago: University of Chicago Press.

——— (1997). *Huxley: From Devil's Disciple to Evolution's High Priest.* Reading, MA: Addison-Wesley.

Desmond, Adrian, and Moore, James (1992). *Darwin: The Life of a Tormented Evolutionist.* New York: Warner Books.

Dewey, John (1910). *The Influence of Darwin on Philosophy and Other Essays.* New York: H. Holt & Co.

Di Gregario, Mario A. (1982). "The Dinosaur Connection: A Reinterpretation of T.H. Huxley's Evolutionary View." *Journal of the History of Biology* 15, 397–418.

Di Gregario, Mario A., and Gill, Nick W., eds. (1990). *Charles Darwin's Marginalia.* Volume 1. New York: Garland Publishing.

Dobzhansky, Theodosius (1937). *Genetics and the Origin of Species.* New York: Columbia University Press.

——— (1973). "Nothing in Biology Makes Sense Except in the Light of Evolution." *American Biology Teacher* 35, 125–129.

Dupré, John (1981). "Natural Kinds and Biological Taxa." *The Philosophical Review* 90, 66–90.

Dupree, A. Hunter (1959). *Asa Gray: American Botanist, Friend of Darwin.* Cambridge: Harvard University Press.

Ehrlich, Paul R. (2002). *Human Natures: Genes, Cultures, and the Human Prospect.* London: Penguin Books.

Ehrlich, Paul R., and Raven, Peter H. (1969). "Differentiation of Populations." *Science* 165, 1228–1232. Reprinted in Ereshefsky (1992a, 57–67).

Eiseley, Loren (1958). *Darwin's Century: Evolution and the Men Who Discovered It.* Garden City, NY: Doubleday.

Eldredge, Niles (1985). *Time Frames.* Princeton: Princeton University Press.

Eldredge, Niles, and Cracraft, Joel (1980). *Phylogenetic Patterns and the Evolutionary Process.* New York: Columbia University Press.

Eldredge, Niles, and Gould, Stephen Jay (1972). "Punctuated Equilibria: An Alternative to Phyletic Gradualism." In T.J.M. Schopf, ed. (1972, 82–115). *Models in Paleobiology.* San Francisco: Cooper and Company.

Ellegård, Alvar (1958). *Darwin and the General Reader: The Reception of Darwin's Theory of Evolution in the British Periodical Press, 1859–1872.* Stockholm: Almqvist and Wiksell.

Endersby, Jim (2003). "Escaping Darwin's Shadow." *Journal of the History of Biology* 36, 385–403.

Endler, John A. (1986). *Natural Selection in the Wild.* Princeton: Princeton University Press.

——— (1989). "Conceptual and Other Problems in Speciation." In Otte and Endler (1989, 625–648).

Ereshefsky, Marc, ed. (1992a). *The Units of Evolution: Essays on the Nature of Species.* Cambridge: MIT Press.

——— (1992b). "Eliminative Pluralism." *Philosophy of Science* 59, 671–690.

Erwin, Douglas H., and Anstey, Robert L., eds. (1995). *New Approaches to Speciation in the Fossil Record.* New York. Columbia University Press.

Farber, Paul L. (1972). "Buffon and the Concept of Species." *Journal of the History of Biology* 5, 259–284.

Fawcett, Henry (1860). "A Popular Exposition of Mr. Darwin on the Origin of Species." *Macmillan's Magazine* 3, 81–92.

Fodor, Jerry A. (1975). *The Language of Thought*. New York: Crowell. Reprinted (1980). Cambridge: Harvard University Press.

Foucault, Michel (1971). *The Order of Things: An Archaeology of the Human Sciences*. New York: Random House.

Fox, Anthony (1995). *Linguistic Reconstruction: An Introduction to Theory and Method*. Oxford: Oxford University Press.

Freeman, Scott, and Herron, Jon C. (1998). *Evolutionary Analysis*. Upper Saddle River, NJ: Prentice Hall.

Frost, Darrel L., and Wright, John W. (1988). "The Taxonomy of Uniparental Species, with Special Reference to Parthenogenetic *Cnemidophorus* (Squamata: Teiidae)." *Systematic Zoology* 37, 200–209.

Futuyma, Douglas J. (1986). *Evolutionary Biology*. 2nd ed. Sunderland, MA: Sinauer Associates.

——— (1998). *Evolutionary Biology*. 3rd ed. Sunderland, MA: Sinauer Associates.

Gayon, Jean (2003). "From Darwin to Today in Evolutionary Biology." In Hodge and Radick (2003, 240–264).

Ghiselin, Michael T. (1966). "On Psychologism in the Logic of Taxonomic Controversies." *Systematic Zoology* 15, 207–215.

——— (1969). *The Triumph of the Darwinian Method*. Berkeley: University of California Press.

——— (1974). "A Radical Solution to the Species Problem." *Systematic Zoology* 23, 536–544.

——— (1987). "Species Concepts, Individuality, and Objectivity." *Biology & Philosophy* 2, 127–143.

——— (1989). "Individuality, History and Laws of Nature in Biology." In Ruse (1989, 53–66).

——— (1997). *Metaphysics and the Origin of Species*. Albany: State University of New York Press.

Ghiselin, Michael, and Jaffe, Linda (1973). "Phylogenetic Classification in Darwin's *Monograph on the Sub-Class Cirripedia*." *Systematic Zoology* 22, 132–140.

Gorovitz, Samuel, *et al.* (1979). *Philosophical Analysis: An Introduction to Its Language and Techniques*. 3rd ed. New York: Random House.

Gould, Stephen Jay (1970). "Dollo on Dollo's Law: Irreversibility and the Status of Evolutionary Laws." *Journal of the History of Biology* 3, 189–212.

——— (1980). *The Panda's Thumb*. New York: W.W. Norton.

——— (1981). *The Mismeasure of Man*. New York: W.W. Norton.

——— (1982). "Darwinism and the Expansion of Evolutionary Theory." *Science* 216, 380–387. Reprinted in Michael Ruse, ed. (1989, 100–117). *Philosophy of Biology*. New York: Macmillan Publishing Company.

———— (1989). *Wonderful Life: The Burgess Shale and the Nature of History.* London: Hutchinson Radius.

Grant, Verne (1957). "The Plant Species in Theory and Practice." In Mayr (1957b, 39–80).

Gray, Asa (1860a). "Review of Darwin's Theory on the Origin of Species by means of Natural Selection." *American Journal of Science and Arts* 29 (2nd ser.), 153–184.

———— (1860b). "Darwin on the Origin of Species." *Atlantic Monthly* 6, 109–116, 229–239.

———— (1860c). "Darwin and His Reviewers." *Atlantic Monthly* 6, 406–425.

Grene, Marjorie, and Depew, David. (2004). *The Philosophy of Biology: An Episodic History.* Cambridge: Cambridge University Press.

Gruber, Howard E. (1981). *Darwin on Man: A Psychological Study of Scientific Creativity.* 2nd ed. Chicago: University of Chicago Press.

Hacking, Ian (1983). *Representing and Intervening: Introductory Topics in the Philosophy of Natural Science.* Cambridge: Cambridge University Press.

Hamilton, Alexander (1820). "Review of Bopp, *Conjugations System.*" *Edinburgh Review* 33, 431.

Harman, Oren Solomon (2003). "C.D. Darlington and the British and American Reaction to Lysenko and the Soviet Conception of Science." *Journal of the History of Biology* 36, 309–352.

Harris, Randy Allen (1993). *The Linguistics Wars.* Oxford: Oxford University Press.

Hattiangadi, J.N. (1987). *How Is Language Possible?: Philosophical Reflections on the Evolution of Language and Knowledge.* La Salle, IL: Open Court.

Haughton, Samuel (1860). "Biogenesis." *Natural History Review* 7, 23–32.

Haynes, Robert H. (1987). "The 'Purpose' of Chance in Light of the Physical Basis of Evolution." In John M. Robson, ed. (1987, 1–31). *Origin and Evolution of the Universe.* Kingston and Montreal, CA: McGill-Queen's University Press.

Hennig, Willi (1966). *Phylogenetic Systematics.* D. Dwight Davis and Rainer Zangerl, trans. Urbana: University of Illinois Press.

Herschel, John F.W. (1830). *A Preliminary Discourse on the Study of Natural Philosophy.* London: Longman, Rees, Orme, Brown, Green, and J. Taylor.

Hodge, M.J.S. (Jon) (1977). "The Structure and Strategy of Darwin's 'Long Argument.'" *British Journal for the History of Science* 10, 237–246.

———— (1987). "Darwin, Species and the Theory of Natural Selection." In Roger and Fischer (1987, 227–252).

———— (1989). "Darwin's Theory and Darwin's Argument." In Ruse (1989, 163–182).

———— (2000). "Knowing about Evolution: Darwin and His Theory of Natural Selection." In Creath and Maienschein (2000, 27–47).

Hodge, Jonathan, and Radick, Gregory, eds. (2003). *The Cambridge Companion to Darwin.* Cambridge: Cambridge University Press.

Hoenigswald, Henry M., and Wiener, Linda F., eds. (1987). *Biological Metaphor and Cladistic Classification: An Interdisciplinary Perspective.* Philadelphia: University of Pennsylvania Press.

Hooker, Joseph Dalton (1859). "Review of the *Origin of Species*." *The Gardeners' Chronicle and Agricultural Gazette* Dec. 31, 1051–1052.

Hopkins, William (1860). "Physical Theories of the Phenomena of Life." *Fraser's Magazine* 61, 739–752; 62, 74–90.

Howard, Daniel J., and Berlocher, Stewart H., eds. (1998). *Endless Forms: Species and Speciation*. New York: Oxford University Press.

Howard, Jonathan (1989). *Darwin*. Oxford: Oxford University Press.

Hoyningen-Huene, Paul (1993). *Reconstructing Scientific Revolutions: Thomas S. Kuhn's Philosophy of Science*. Alexander T. Levine, trans. Chicago: Chicago University Press.

Hull, David L. (1965). "The Effect of Essentialism on Taxonomy: Two Thousand Years of Stasis." *British Journal for the Philosophy of Science* 15, 314–326; 16, 1–18.

—— (1978). "A Matter of Individuality." *Philosophy of Science* 45, 335–360.

—— (1988). *Science as a Process*. Chicago: University of Chicago Press.

—— (2000). "The Professionalization of Science Studies: Cutting Some Slack." *Biology & Philosophy* 15, 61–91.

Hull, David L., Tessner, Peter D., and Diamond, Arthur M. (1978). "Planck's Principle: Do Younger Scientists Accept New Scientific Ideas with Greater Alacrity than Older Scientists?" *Science* 202, 717–723. Reprinted in Hull (1989, 43–61). *The Metaphysics of Evolution*. Albany: State University of New York Press.

Huxley, Julian S. (1912). *The Individual in the Animal Kingdom*. Cambridge: Cambridge University Press.

Huxley, Thomas H. (1859a). "On the Persistent Types of Animal Life." *Proceedings of the Royal Institution of Great Britain 1858–1862* 3, 151–153.

—— (1859b). "Time and Life: Mr. Darwin's 'Origin of Species.'" *Macmillan's Magazine* 1, 142–148.

—— (1859c). "Darwin on the Origin of Species." *The Times* Dec. 26, 8.

—— (1860a). "On Species and Races, and their Origin." *Proceedings of the Royal Institution of Great Britain 1858–1862* 3, 195–200.

—— (1860b). "On the Origin of Species by means of Natural Selection." *Westminster Review* 17, 541–570.

Keller, Evelyn Fox, and Lloyd, Elisabeth A., eds. (1992). *Keywords in Evolutionary Biology*. Cambridge: Harvard University Press.

Kitcher, Philip (1978). "Theories, Theorists and Theoretical Change." *The Philosophical Review* 87, 519–547.

—— (1993). *The Advancement of Science: Science without Legend, Objectivity without Illusions*. New York: Oxford University Press.

Kohn, David, ed. (1985a). *The Darwinian Heritage*. Princeton: Princeton University Press.

—— (1985b). "Darwin's Principle of Divergence as Internal Dialogue." In Kohn (1985a, 245–257).

Kondrashov, Alexey S., Yampolsky, Lev Yu, and Shabalina, Svetlana A. (1998). "On the Sympatric Origin of Species by Means of Natural Selection." In Howard and Berlocher (1998, 90–98).

Kottler, Malcolm Jay (1978). "Charles Darwin's Biological Species Concept and Theory of Geographic Speciation: the Transmutation Notebooks." *Annals of Science* 35, 275–297.

—— (1985). "Charles Darwin and Alfred Russel Wallace: Two Decades of Debate Over Natural Selection." In Kohn (1985a, 367–432).

Kuhn, Thomas S. (1957). *The Copernican Revolution.* Cambridge: Harvard University Press.

—— (1970). *The Structure of Scientific Revolutions.* 2nd ed. Chicago: University of Chicago Press.

—— (1977). *The Essential Tension: Selected Studies in Scientific Tradition and Change.* Chicago: University of Chicago Press.

—— (1993). "Afterwords." In Paul Horwich, ed. (1993, 311–341). *World Changes: Thomas Kuhn and the Nature of Science.* Cambridge: MIT Press.

Lakatos, Imre (1970). "Falsification and the Methodology of Scientific Research Programmes." In Lakatos and Alan Musgrave, eds. (1970, 91–196). *Criticism and the Growth of Knowledge.* Cambridge: Cambridge University Press.

Lamarck, Jean Baptiste de (1809). *Philosophie Zoologique.* Paris: Dentu. Hugh Elliot, trans. (1914). *Zoological Philosophy.* New York: Macmillan.

Lambert, David M., and Spencer, Hamish G., eds. (1995). *Speciation and the Recognition Concept: Theory and Applications.* Baltimore: Johns Hopkins.

LaPorte, Joseph (2004). *Natural Kinds and Conceptual Change.* Cambridge: Cambridge University Press.

Laporte, Léo F. (1994). "Simpson on Species." *Journal of the History of Biology* 27, 141–159.

Larson, James L. (1968). "The Species Concept of Linnaeus." *Isis* 59, 291–299.

Lass, Roger (1997). *Historical Linguistics and Language Change.* Cambridge: Cambridge University Press.

Leifchild, John R. (1859). "On the Origin of Species by Means of Natural Selection." *The Athenæum* Nov. 19, 659–660.

Leikola, Anto (1987). "The Development of the Species Concept in the Thinking of Linnaeus." In Roger and Fischer (1987, 45–59).

Levin, Donald A. (1979). "The Nature of Plant Species." *Science* 204, 381–204.

—— (1983). "Polyploidy and Novelty in Flowering Plants." *American Naturalist* 122, 1–25.

Lewes, George Henry (1860). "Studies in Animal Life." *Cornhill Magazine* 4, 438–447.

Lightman, Bernard, ed. (1997). *Victorian Science in Context.* Chicago: University of Chicago Press.

Lipton, Peter. (1990). "Contrastive Explanation." In Dudley Knowles, ed. (1990, 247–266). *Explanation and its Limits.* Cambridge: Cambridge University Press.

Lovejoy, Arthur O. (1936). *The Great Chain of Being: A Study of the History of an Idea.* Cambridge: Harvard University Press.

——— (1959). "Buffon and the Problem of Species." In Bentley Glass, *et al.*, eds. (1959, 84–113). *Forerunners of Darwin: 1745–1859.* Baltimore: Johns Hopkins.

Luckow, Melissa (1995). "Species Concepts: Assumptions, Methods, and Applications." *Systematic Botany* 20, 589–605.

Lyell, Charles (1830). *Principles of Geology.* Volume 1. London: John Murray.

——— (1832). *Principles of Geology.* Volume 2. London: John Murray.

——— (1833). *Principles of Geology.* Volume 3. London: John Murray.

——— (1838). *Elements of Geology.* London: John Murray.

——— (1863). *The Geological Evidences of the Antiquity of Man, with Remarks on Theories of the Origin of Species by Variation.* London: John Murray.

Lyons, Sherrie (1999). "In Search of Huxley the Scientist." *Biology & Philosophy* 14, 585–591.

Mallet, James (1995). "A Species Definition for the Modern Synthesis." *Trends in Ecology and Evolution* 10, 294–299.

Masters, Maxwell T. (1860). "On the Relation between the Abnormal and Normal Formations in Plants." *Notices of the Proceedings at the Meetings of the Members of the Royal Institution of Great Britain (1858–1862)* 3, 223–227.

Mauskopf, Seymour H. (1992). "Prize Announcements." *Isis* 83, 278–279.

Maynard Smith, John (1975). *The Theory of Evolution.* Harmondsworth, UK: Penguin Books. Reprinted (1993). Cambridge: Cambridge University Press.

Mayr, Ernst (1942). *Systematics and the Origin of Species.* New York: Columbia University Press.

——— (1957a). "Species Concepts and Definitions." In Mayr (1957b, 1–22).

———, ed. (1957b). *The Species Problem.* Washington, DC: American Association for the Advancement of Science.

——— (1970). *Populations, Species, and Evolution.* Cambridge: Harvard University Press.

——— (1976). *Evolution and the Diversity of Life.* Cambridge: Harvard University Press.

——— (1982). *The Growth of Biological Thought.* Cambridge: Harvard University Press.

——— (1985). "Darwin's Five Theories of Evolution." In Kohn (1985a, 755–772).

——— (1987). "The Ontological Status of Species: Scientific Progress and Philosophical Terminology." *Biology & Philosophy* 2, 145–166.

——— (1988). *Toward a New Philosophy of Biology.* Cambridge: Harvard University Press.

——— (1991). *One Long Argument: Charles Darwin and the Genesis of Modern Evolutionary Thought.* Cambridge: Harvard University Press.

——— (1992). "A Local Flora and the Biological Species Concept." *American Journal of Botany* 79, 222–238.

Mayr, Ernst, and Short, Lester L. (1970). *Species Taxa of North American Birds.* Cambridge: The Nuttall Ornithological Club.

McDade, Lucinda A. (1995). "Species Concepts and Problems in Practice: Insight from Botanical Monographs." *Systematic Botany* 20, 606–622.

McOuat, Gordon R. (1996). "Species, Rules and Meaning: The Politics of Language and the Ends of Definitions in 19th Century Natural History." *Studies in History and Philosophy of Science* 27, 473–519.

———— (2000). "Networks, Hybrids and Forms of Life." *Annals of Science* 57, 189–195.

———— (2001). "Cataloguing Power: Delineating 'Competent Naturalists' and the Meaning of Species in the British Museum." *British Journal for the History of Science* 34, 1–28.

Mishler, Brent D., and Donoghue, Michael J. (1982). "Species Concepts: A Case for Pluralism." *Systematic Zoology* 31, 491–503.

Moore, James (1996). "Metabiographical Reflections on Charles Darwin." In Shortland and Yeo (1996, 267–281).

———— (1997). "Wallace's Malthusian Moment: The Common Context Revisited." In Lightman (1997, 290–311).

Morgan, T.H. (1903). *Evolution and Adaptation*. New York: Macmillan.

Murray, Andrew (1859). "On Mr. Darwin's Theory of the Origin of Species." *Proceedings of the Royal Society of Edinburgh (1857–1862)* 4, 274–291.

Nagel, Ernest (1961). *The Structure of Science*. New York: Harcourt, Brace & World.

O'Hara, Robert J. (1993). "Systematic Generalization, Historical Fate, and the Species Problem." *Systematic Biology* 42, 231–246.

Ospovat, Dov (1981). *The Development of Darwin's Theory: Natural History, Natural Theology, and Natural Selection, 1838–1859*. Cambridge: Cambridge University Press.

Otte, Daniel, and Endler, John A., eds. (1989). *Speciation and Its Consequences*. Sunderland, MA: Sinauer Associates.

Owen, Richard (1860). "On the Origin of Species by means of Natural Selection." *Edinburgh Review* 3, 487–532.

Padian, Kevin (1999). "Charles Darwin's Views of Classification in Theory and Practice." *Systematic Biology* 48, 352–364.

Parshall, Karen Hunger (1982). "Varieties As Incipient Species: Darwin's Numerical Analysis." *Journal of the History of Biology* 15, 191–214.

Paterson, H.E.H. (1985). "The Recognition Concept of Species." In E.S. Vrba, ed. (1985, 21–29). *Species and Speciation*. Pretoria, South Africa: Transvaal Museum.

Percival, W. Keith (1987). "Biological Analogy in the Study of Languages Before the Advent of Comparative Grammar." In Hoenigswald and Wiener (1987, 3–38).

Poulton, E.B. (1903). "What is a Species?" *Royal Entomological Society of London, Transactions (Appendix)* 51, 77–116.

Prichard, James Cowles (1813). *Researches into the Physical History of Man*. London: John and Arthur Arch.

Putnam, Hilary (1973). "Explanation and Reference." In Putnam, ed. (1975, 196–214). *Mind, Language and Reality*. Cambridge: Cambridge University Press.

Quine, W.V.O. (1986). *Philosophy of Logic.* 2nd ed. Cambridge: Harvard University Press.

Radick, Gregory (2003). "Is the Theory of Natural Selection Independent of Its History?" In Hodge and Radick (2003, 143–167).

Ramsbottom, J. (1938). "Linnaeus and the Species Concept." *Proceedings of the Linnean Society of London* May 24, 192–219.

Raup, David M. (1991). *Extinction: Bad Genes or Bad Luck?* New York: W.W. Norton & Company.

Richards, Robert J. (1992). *The Meaning of Evolution: The Morphological Construction and Ideological Reconstruction of Darwin's Theory.* Chicago: University of Chicago Press.

——— (2002). *The Romantic Conception of Life: Science and Philosophy in the Age of Goethe.* Chicago: University of Chicago Press.

——— (2004). "Michael Ruse's Design for Living." *Journal of the History of Biology* 37, 25–38.

Ridley, Mark (1989). "The Cladistic Solution to the Species Problem." *Biology & Philosophy* 4, 1–16.

——— (1993). *Evolution.* Boston: Blackwell Scientific Publications.

Rieppel, Olivier (1986). "Species Are Individuals: A Review and Critique of the Argument." In Max K. Hecht, *et al.*, eds. (1986, 283–387). *Evolutionary Biology.* Volume 20. New York: Plenum Press.

Ritvo, Harriet (1997). "Zoological Nomenclature and the Empire of Victorian Science." In Lightman (1997, 334–353).

Roger, Jacques, and Fischer, Jean-Louis, eds. (1987). *Histoire du Concept d'Espèce dans les Sciences de la Vie.* Paris: Fondation Singer-Polignac.

Rosen, David (1978). "The Importance of Cryptic Species and Specific Identifications as Related to Biological Control." *Biosystematics in Agriculture: Beltsville Symposia in Agricultural Research* 2, 23–35.

Rosen, Donn E. (1979). "Fishes from the Uplands and Intermontane Basins of Guatemala: Revisionary Studies and Comparative Geography." *Bulletin of the American Museum of Natural History* 162, 267–376.

Ruse, Michael (1975). "Darwin's Debt to Philosophy: An Examination of the Influence of the Philosophical Ideas of John F.W. Herschel and William Whewell on the Development of Charles Darwin's Theory of Evolution." *Studies in the History and Philosophy of Science* 6, 159–181.

——— (1979). *The Darwinian Revolution.* Chicago: University of Chicago Press.

——— (1987). "Biological Species: Natural Kinds, Individuals, or What?" *British Journal for the Philosophy of Science* 38, 225–242. Reprinted in Ereshefsky (1992a, 343–363).

———, ed. (1989). *What the Philosophy of Biology Is.* Dordrecht, Netherlands: Kluwer.

——— (1993a). "Were Owen and Darwin *Naturphilosophen?*" *Annals of Science* 50, 383–388.

——— (1993b). "Will the Real Charles Darwin Please Stand Up?" *Quarterly Review of Biology* 68, 225–231.

———— (2000). "Darwin and the Philosophers: Epistemological Factors in the Development and Reception of the Theory of the *Origin of Species*." In Creath and Maienschein (2000, 3–26).

———— (2004). "The Romantic Conception of Robert J. Richards." *Journal of the History of Biology* 37, 3–23.

Salthe, Stanley N. (1985). *Evolving Hierarchical Systems*. New York: Columbia University Press.

Schwartz, Stephen P., ed. (1977). *Naming, Necessity, and Natural Kinds*. Ithaca, NY: Cornell University Press.

Schweber, Silvan S. (1985). "The Wider British Context in Darwin's Theorizing." In Kohn (1985a, 35–69).

Secord, James A. (1989). "Behind the Veil: Robert Chambers and *Vestiges*." In James R. Moore, ed. (1989). *History, Humanity and Evolution: Essays for John C. Greene*. Cambridge: Cambridge University Press.

Sedgwick, Adam (1860). "Objections to Mr. Darwin's Theory of the Origin of Species." *The Spectator* Mar. 24, 285–286; Apr. 7, 334–335.

Shortland, Michael, and Yeo, Richard, eds. (1996). *Telling Lives in Science: Essays on Scientific Biography*. Cambridge: Cambridge University Press.

Simpson, George Gaylord (1961). *Principles of Animal Taxonomy*. New York: Columbia University Press.

Simpson, Richard (1860). "Darwin on the Origin of Species." *Rambler* 2, 361–376.

Smith, Neil (1999). *Chomsky: Ideas and Ideals*. Cambridge: Cambridge University Press.

Sober, Elliott (1993). *Philosophy of Biology*. Boulder: Westview Press.

Sonea, Sorin, and Panisset, Maurice (1983). *A New Bacteriology*. Boston: Jones and Bartlett.

Stamos, David N. (1996). "Was Darwin Really a Species Nominalist?" *Journal of the History of Biology* 29, 127–144.

———— (1998). "Buffon, Darwin, and the Non-Individuality of Species—A Reply to Jean Gayon" *Biology & Philosophy* 13, 443–470.

———— (1999). "Darwin's Species Category Realism." *History and Philosophy of the Life Sciences* 21, 21–70.

———— (2000). "Book Review of Stephen G. Alter's *Darwinism and the Linguistic Image: Language, Race, and Natural Theology in the Nineteenth Century*." *Annals of Science* 57, 319–321.

———— (2002). "Species, Languages, and the Horizontal/Vertical Distinction." *Biology & Philosophy* 17, 171–198.

———— (2003). *The Species Problem: Biological Species, Ontology, and the Metaphysics of Biology*. Lanham, MD: Lexington Books.

———— (2005). "Pre-Darwinian Taxonomy and Essentialism—A Reply to Mary Winsor." *Biology & Philosophy* 20, 79–96.

———— (2006). "Popper, Laws, and the Exclusion of Biology from Genuine Science." manuscript in review.

——— (forthcoming). *Evolution and the Big Questions: Sex, Race, Religion, and Other Matters*. Blackwell Publishing.

Stauffer, R.C., ed. (1975). *Charles Darwin's Natural Selection, Being the Second Part of His Big Species Book Written From 1856 to 1858*. Cambridge: Cambridge University Press.

Stevens, Peter F. (1980). "Evolutionary Polarity of Character States." *Annual Review of Ecology and Systematics* 11, 333–358.

——— (1984). "Haüy and A.-P. Candolle: Crystallography, Botanical Systematics, and Comparative Morphology, 1780–1840." *Journal of the History of Biology* 17, 49–82.

——— (1992). "Species: Historical Perspectives." In Keller and Lloyd (1992, 302–311).

——— (1994). *The Development of Biological Systematics: Antoine-Laurent de Jussieu, Nature, and the Natural System*. New York: Columbia University Press.

——— (1997). "J.D. Hooker, George Bentham, Asa Gray and Ferdinand Mueller on Species Limits in Theory and Practice: A Mid-Nineteenth-Century Debate and Its Repercussions." *Historical Records of Australian Science* 11, 345–370.

Strickland, Hugh E. (1837). "Rules for Zoological Nomenclature." *Magazine of Natural History* 1, 173–176.

——— (1842). "Report of a Committee Appointed 'to Consider of the Rules by which the Nomenclature of Zoology may be Established on a Uniform and Permanent Basis.'" *British Association for the Advancement of Science (London)*, 105–121.

Sulloway, Frank J. (1979). "Geographic Isolation in Darwin's Thinking: The Vicissitudes of a Crucial Idea." In William Coleman and Camille Limoges, eds. (1979, 23–65). *Studies in History of Biology*. Volume 3. Baltimore: Johns Hopkins.

——— (1982). "Darwin's Conversion: The *Beagle* Voyage and Its Aftermath." *Journal of the History of Biology* 15, 325–396.

——— (1996). *Born to Rebel: Birth Order, Family Dynamics, and Creative Lives*. New York: Pantheon Books.

Taub, Liba (1993). "Evolutionary Ideas and 'Empirical' Methods: The Analogy Between Language and Species in Works by Lyell and Schleicher." *British Journal for the History of Science* 26, 171–193.

Templeton, Alan R. (1989). "The Meaning of Species and Speciation: A Genetic Perspective." In Otte and Endler (1989, 1–27).

Thompson, Paul (1989). *The Structure of Biological Theories*. Albany: State University of New York Press.

———, ed. (1995). *Issues in Evolutionary Ethics*. Albany: State University of New York Press.

Trask, R.L. (1996). *Historical Linguistics*. London: Arnold.

Van Lalen, Leigh (1973). "A New Evolutionary Law." *Evolutionary Theory* 1, 1–30.

——— (1976). "Ecological Species, Multispecies, and Oaks." *Taxon* 25, 233–239. Reprinted in Ereshefsky (1992a, 69–78).

Wallace, Alfred Russel (1855). "On the Law Which Has Regulated the Introduction of New Species." *Annals and Magazine of Natural History* 16 (2nd ser.), 184–196.

———— (1859). "On the Tendency of Varieties to depart indefinitely from the Original Type." *Journal of the Proceedings of the Linnean Society (Zoology)* 3, 53–62. Reprinted in Barrett (1977 II, 10–19).

Watson, Hewett Cottrell (1843). "Remarks on the Distinction of Species in Nature, and in Books; preliminary to the notice of some variations and transitions of character, observed in the native plants of Britain." *London Journal of Botany* 2, 613–622.

———— (1845a). "On the Theory of 'Progressive Development,' applied in explanation of the Origin and Transmutation of Species." *The Phytologist* 2, 108–113, 140–147, 161–168, 225–228.

———— (1845b). "Report of an experiment which bears upon the specific identity of the Cowslip and Primrose." *The Phytologist* 2, 217–219.

———— (1859). *Cybele Britannica*. Volume 4. London: Longman.

Wayne, Robert K., and Gittleman, John L. (1995). "The Problematic Red Wolf." *Scientific American* 273, 36–39.

Wedgwood, Hensleigh (1833). "Grimm's *Deutche Grammatik*." *Quarterly Review* 50, 169–189.

Weinberg, Steven (1992). *Dreams of a Final Theory: The Scientist's Search for the Ultimate Laws of Nature*. New York: Pantheon Books.

Weinert, Friedel, ed. (1995). *Laws of Nature: Essays on the Philosophical, Scientific and Historical Dimensions*. Berlin: Walter de Gruyter.

Wells, Rulon S. (1987). "The Life and Growth of Language: Metaphors in Biology and Linguistics." In Hoenigswald and Wiener (1987, 39–80).

West-Eberhard, Mary Jane (1992). "Adaptation: Current Usages." In Keller and Lloyd (1992, 13–18).

Wheeler, Quentin D., and Meier, Rudolf, eds. (2000). *Species Concepts and Phylogenetic Theory*. New York: Columbia University Press.

Whewell, William (1837). *History of the Inductive Sciences*. Three volumes. London: Parker.

———— (1840). *Philosophy of the Inductive Sciences*. Two volumes. London: Parker.

Wilberforce, Samuel (1860). "Darwin's *Origin of Species*." *Quarterly Review* 108, 225–264.

Wiley, E.O. (1981). *Phylogenetics: The Theory and Practice of Phylogenetic Systematics*. New York: John Wiley & Sons.

Wilson, Edward O. (1992). *The Diversity of Life*. Cambridge: Harvard University Press.

Wilson, Leonard G., ed. (1970). *Sir Charles Lyell's Scientific Journals on the Species Question*. New Haven: Yale University Press.

Wilson, Robert A., ed. (1999). *Species: New Interdisciplinary Essays*. Cambridge: MIT Press.

Winsor, Mary Pickard (1979). "Louis Agassiz and the Species Question." In William Coleman and Camille Limoges, eds. (1979, 89–117). *Studies in History of Biology*. Volume 3. Baltimore: Johns Hopkins.

———— (2001). "The Practitioner of Science: Everyone Her Own Historian." *Journal of the History of Biology* 34, 229–245.

———— (2003). "Non-Essentialist Methods in Pre-Darwinian Taxonomy." *Biology & Philosophy* 18, 387–400.

Wollaston, Thomas Vernon (1856). *On the Variation of Species, With Especial Reference to the Insecta*. London: John Van Voorst.

———— (1860). "On the Origin of Species by means of Natural Selection." *Annals and Magazine of Natural History* 5 (3rd ser.), 132–143.

Index